ARMY Trivia

EDWARD J. BURKE

Quinlan Press
Boston, Massachusetts

Published by Quinlan Press, Inc.
131 Beverly Street
Boston, MA 02114
(617) 227-4870

Cover design by Joan McLaughlin

Library of Congress Catalog Card Number
85-62498
ISBN 0-933341-14-8

First printing August 1985
Second printing September 1985

The views expressed in this book are those of the author and do not reflect the official policy or position of the Department of the Army, Department of Defense, or the US Government.

To my father and brother.

Ed Burke is a 1961 graduate of Boston College and a colonel in the United States Army Reserve. While at Boston College, he majored in English, played hockey, and was commissioned in the infantry through the ROTC program. His Army service includes periods in the Active Army, Army Reserve and AGR program. While on active duty, he served two tours of duty in Vietnam, was wounded, and was decorated for bravery. A graduate of the US Army War College, he is currently assigned to the Pentagon in Washington, D.C.

FOREWORD

This book celebrates the history of the US Army in a contemporary, popular form. It is designed for the broadest possible audience. To the extent that it informs, educates, promotes study and deepens appreciation for this most unique of American institutions, I will have done my duty to my comrades.

I would like to acknowledge the following people:

For the idea, and much more—my lifelong friend Henry Quinlan.

For patiently typing from thousands of scribbled 3x5 cards—Debbie Mulcahy and Anna Silsbee.

For meticulously editing and patiently enduring another novice—Sandy Bielawa.

For aggressively and imaginatively "pushing" this book—Kathy Joyce.

For saving me from "Dopey Wise" and many other perils—Dave Johnson.

For keeping somewhat quieter than usual—Maryalice, Sean and Michael.

For everything else—Sally.

Table of Contents

CONFLICT

1. What was important about the Battle of Five Forks, Virginia, which took place on April 1, 1865?

2. On what date was the "shot heard around the world" fired?

3. Where was the shot fired?

4. Where was the famous Battle of Bunker Hill in Charlestown, Massachusetts actually conducted?

5. What event occurred on Sullivan's Island in Charleston, South Carolina on June 28, 1776?

6. What famous event occurred on December 25, 1776?

7. Where did George Washington's army encamp during the winter of 1777-78?

8. Where did the Continental Army victory that ended the Revolutionary War take place?

9. At what battle during the War of 1812 did a surprised British commander exclaim, "Those are regulars, by God!"?

10. What war began with the ambush of an Army column called the Dade Massacre?

11. What war began with an attack on the forces commanded by Brigadier General Zachary Taylor?

12. Name the last Indian battle of consequence, which occurred in 1890.

13. Name the domestic turmoil that occurred in 1857-58 and required the commitment of the Army in a western state.

14. What war began with the sinking of the battleship **Maine**?

15. Who was victorious at the Battle of Cowpens on January 17, 1781?

16. Who was "Old Abe" (not President Lincoln)?

17. What was the greatest single disaster to the American cause during the Revolutionary War?

18. To what did the slogan "Remember the River Raisin" refer?

19. What was unusual about the Battle of Petersburg on July 31, 1864?

20. In September of 1862, some 13,000 Union and 8,000 Confederate troops

died, yet tactically the battle was a draw. Name the battle.

21. What Civil War offensive action was composed of forty-seven assault regiments totalling 15,000 men?

22. Under what circumstances did the first casualty of the Civil War occur?

23. Name the Civil War unit commanded by John Singleton Mosby.

24. During the Mexican War at the Battle of Buena Vista, Zachary Taylor rode up to an artillery battery and said to its commander, "A little more grape, Captain ____________."

25. Name the single horse that survived the Battle of Little Big Horn and for the rest of his days appeared saddled, but riderless, at all 7th Cavalry parades.

26. What was George Armstrong Custer's rank when he was killed?

27. From what tribe was the Indian chief Geronimo?

28. Who commanded the punitive expedition sent to Mexico in 1916?

29. During what conflict did "Rogers' Rangers" fight?

30. In what war was the helicopter first used on any scale?

31. What weapon was used extensively in WWI, but then prohibited by convention?

32. Name the Army battery that fired the first American shot in WWI.

33. At which battle did Confederate General Pierre G.T. Beauregard conduct one of the biggest ruses of the Civil War?

34. Who received the surrender of the confederate forces at Appomattox Courthouse?

35. What two countries assisted America during the American Revolution?

36. To what did the term "Flying Camp," used during the defense of New York in the Revolutionary War, refer?

37. What Indian tribe attacked Brigadier General William Henry Harrison at the Battle of Tippecanoe?

38. Which Union general ordered the suicidal uphill assault at the Battle of Fredericksburg?

39. Who presented Savanah, Georgia to President Abraham Lincoln as a Christmas present in 1864?

40. Which US Army corps liberated Paris during WWII and was honored by the French people as conquering heroes?

41. Where did the Japanese conduct their second major air raid the day after Pearl Harbor?

42. In which US war did the greatest number of American casualties occur?

43. Name the site in North Africa where American and British armored forces defeated Rommel.

44. Name the German puppet government in southern France which opposed the Army in WWII.

45. Name the system of French fortification on the eastern French border extending from Belgium to Switzerland that was outflanked by the Germans in 1941.

46. Name the emperor of Japan during WWII who later renounced his imperial divinity during US Army occupation.

47. What was the German term used to describe a rapid, violent, overwhelming attack?

48. Name the French premiere who was the leader of the Vichy forces against the US Army.

49. Name the famous Solomon Island recaptured by US forces in 1943 that is also the title of a best-selling book.

50. Name the place where MacArthur executed a surprise thrust to the rear of enemy positions in Korea.

51. What was the name of the partially trained South Korean soldiers integrated into US Army units during the Korean War?

52. After being airlifted to Pusan, the 24th Infantry Division had particular difficulty defending against what kind of tanks?

53. In August of 1950, the Army's Korean War strategy consisted of holding a triangle to prevent ejection from the Korean Peninsula. This triangle was described by the cities of Taegu, Pohang and __________.

54. What kind of Chinese Communist shelters—often large enough to accomodate a company or battalion—were employed in the Korean War?

55. After over thirty days of intense fighting in September/October of 1951, at a cost of over 3700 casualties, the 2d Infantry Division was in possession of what terrain feature?

56. During the Korean War, the Chinese Communists made extensive use of subterranean shelters due to their fear of what?

57. What occurred at the towns of Sukchon/Sunchon and Munsan-ni during the Korean War?

58. What was Operation Bluehearts?

59. Name the first American unit committed during the Korean War.

60. What division followed the 1st Marine Division ashore at Inchon?

61. Name the place in North Korea where the second amphibious landing of Army forces occurred.

62. Which was the first infantry division committed to combat during the Korean War?

63. What was the date of the Inchon landing in the Korean War?

64. During the summer of 1952, much of the focus of Korean War activity centered around a hill called "Old Baldy." Name the division which took Old Baldy and successfully defended it against many savage counter-attacks.

65. What was the greatest single mass troop operation in Army history?

66. How many Americans were killed at the Battle of the Alamo on March 6, 1846?

67. Pershing considered it an American victory, yet the Meuse-Argonne resulted in how many American casualties?

68. What did the term CIDG, used during the Vietnam War, stand for?

69. What division was the rear guard unit for X Corps' evacuation at Hungnam in December of 1950?

70. During the Korean War, what was "Task Force Dog"?

71. What was "The MacNamara Line" in Vietnam?

72. Name the division which distinguished itself in the A Shau Valley of South Vietnam.

73. What were "Ruff Puffs"?

74. Name the place where the biggest battle of the Vietnam War occurred.

75. What was the name of the largest US Army operation of the Vietnam War?

76. Which Special Forces group controlled Special Forces operations in the Vietnam War?

77. What is the common name of the last major German offensive of WWII, which occurred in December 1944/January 1945?

78. Name the seven Army divisions that fought in the Vietnam War.

79. Approximately how many Continental Army soldiers were killed in the Revolutionary War?

80. Where did George Armstrong Custer make his "last stand"?

81. What map reference was used as the line of demarcation during the Korean War?

82. Name the eight separate brigades/regiments that fought in the Vietnam War.

83. In WWI, a battalion of war widows from what country served with the Allied forces?

84. Which division reached the Yalu River at Hyesanjin, the most northerly point occupied by US forces during the Korean War?

85. From what port city in North Korea did X Corps conduct its brilliant withdrawal in December of 1950?

86. Name the place where the 23d Infantry won a Presidential Unit Citation for its famous defensive stand during the Korean War.

87. What was Operation Dewey Canyon II, conducted in early 1971?

88. What event occurred on November 21, 1970?

89. During the Vietnam War, what was significant about the "Fish Hook," the "Parrot's Beak," and the "Bulge"?

90. What was "The City" that was discovered during the Cambodian incursion?

91. What was the full-scale assault conducted by the enemy on all major South Vietnamese cities in early 1968 called?

92. What was the name given to the Ninth Infantry Division's unique operations in the Mekong Delta of South Vietnam?

93. What was the most famous song to emanate from WWI?

94. Name the weapon for which General Matthew Ridgeway reinstituted training for combat soldiers during the Korean War.

95. Give the approximate total strength of Communist Chinese who crossed the Yalu River into Korea in October of 1950.

96. On what date was the first atomic bomb dropped?

97. What was the rare but highly publicized Communist Chinese offensive tactic during the Korean War called?

98. What was the name of the tactical period which began with a month-long battle in September of 1951?

99. In the spring of 1961, President John F. Kennedy made a key decision to increase the number of Army advisors in South Vietnam to how many?

100. What geographic feature was the line of demarcation between North and South Vietnam?

101. What was the term used to describe the fluid nature of Army operations in North Africa in WWII?

102. In September, 1944 the US Army's first penetration of the German border occurred at what point?

103. Where was the WWII surrender document for the European theatre signed?

104. What was the only weapon used by the Japanese to kill civilians on the US mainland during WWII?

105. Name the famous bridge over the Rhine at Remagen, luckily captured intact by the US Army.

106. What was the defensive line that guarded the heartland of Germany in WWII called?

107. On June 6, 1944 (D-Day) sixty-one top German leaders were absent from their units. What were they doing?

108. Because of its adverse affect on the German Army, what did Rommel call D-Day?

109. Name the five beaches that comprised the Normandy beachhead.

110. For what was the Santo Tomas Internment Camp famous?

111. What was the first major US Army unit to cross the Seine River in WWII?

112. Name the longest war in US history.

113. Name the American division whose soldiers first set foot on captive Japan.

114. What was the name of the largest battle of the Pacific theatre in WWII?

115. How many Army divisions were involved in the battle for Okinawa in WWII?

116. This name for a prominent terrain feature was common to both the Korean War and the Vietnam War. What was it?

117. What was the American name for South Vietnam's Highway 13, which ran northwest from Saigon?

118. Name the Army units that participated in the combat jump onto the island of Grenada.

119. How long did it take first the Marines and then the Army to clear Japanese forces from the island of Guadalcanal?

120. Where did US and German forces have their first major battle of WWII?

121. What was the Gothic Line?

122. Where was the final land battle of WWII fought?

123. Where was the 82d Airborne Division deployed on April 30, 1965?

124. What was the name given to the area that divided North and South Vietnam along the 17th parallel?

125. Name the valley in the western highlands of Vietnam that was the site of the 1st Cavalry Division's victory over North Vietnamese forces in November, 1965.

126. Name the battalion which parachuted into War Zone C in February of 1967 during the Vietnam War.

127. Name the South Vietnamese village completely evacuated and destroyed during Operation Cedar Falls, about which a book was later written.

128. What name was given to the area of Santo Domingo, initially cleared by the Marines, to which the 82d Airborne Division established a corridor and then linked up with the Marines?

129. The US Army landed on two of the five invasion beaches of Normandy. Name them.

130. During the Battle of Château-Thierry, what American division repulsed the Germans and captured Belleau Wood?

131. What was the "Great War"?

132. Where is "Kettle Hill," and what is its significance?

133. Name the decisive AEF offensive that was a prelude to victory in WWI.

134. What was the name of the first major WWI victory won by the US Army under its own command?

135. Who was victorious at the Battle of the Marne in July, 1918?

136. What was the name of the massive trucking operation that rushed supplies eastward after the Normandy landing?

137. Name the American division that repulsed German efforts to cross the

Marne River during the Battle of Château-Thierry in WWI.

138. Name the exact place where Lee surrendered to Grant on April 9, 1865.

139. What battle was the real turning point of the Civil War?

140. How many nations were allied with the US during the Korean War?

141. How many West Point graduates served in the Confederate Army during the Civil War?

142. What were the Japanese code words that confirmed the attack on Pearl Harbor?

143. Named for their commander, they were Army ground forces in the Battle of Burma. Who were they?

144. What was the famous Japanese assault cry signaling victory or death?

145. How many West Point-educated Confederate generals died during the Civil War?

146. Where did the first US Army airborne assault in WWII take place?

147. Name the famous landmark under which AEF forces marched in a gigantic parade celebrating the end of WWI.

148. The Meuse-Argonne in WWI was nicknamed the ____________________

because of the huge number of casualties suffered there.

149. What was the Lafayette Escadrille?

150. When the Lafayette Escadrille was disbanded, some of its members joined the Army's famous 94th Squadron. What was its nickname?

151. Where were the first three American soldiers to die in WWI killed?

152. Name the place where a massive riot of Communist POW's occurred during the Korean War.

153. When Korean War prisoner lists were released by both sides, the Chinese accounted for how many US prisoners?

154. What was the name of the devices used by the Chinese Communists in Korea to enable convoy movement at night?

155. What was the name of the operation planned by the Army in August of 1951 to paralyze the Communist rear area?

156. How many divisions did the US Army employ in the Korean War?

157. What Army division is credited with the capture of both Field Marshall Gerd von Rundstedt and Reichsmarshal Hermann Goering?

158. Name the "Lost Battalion" of WWII.

159. What was the "Carpetbagger Squadron"?

160. What was the WWII "Galahad Force"?

161. What was the code name for the invasion of Sardinia in WWII?

162. What was the Allied code name for the US Army in WWII?

163. Who was code-named "Duckpin" during WWII?

164. Name WWII's famous "Go for Broke" unit.

165. The construction of what road was one of the most extraordinary achievements of WWII?

166. Name the battle which was the worst US defeat in North Africa in WWII.

167. Allied bombing of what historic abbey in WWII sparked a world-wide wave of protest?

168. Where was "The Hump"?

169. What WWII battle was compared to the Argonne of WWI?

170. Who was the most popular WWII pin-up girl?

171. Name the place where the Allies trapped the German Fifth and Seventh Panzer Armies in a classic pincers movement in WWII.

172. Name the famous tunnel on Corregidor Island.

173. Where, in WWII, were Allied forces a "stranded whale" suffering extremely heavy casualties for four months?

174. Where did the first battle between US and Japanese paratroopers occur?

175. After the Battle of the Hürtgen Forest, what was the 28th Infantry Division called?

176. Name the fort captured in May, 1775 by Colonel Ethan Allen and his "Green Mountain Boys."

177. Who was Lieutenant Harry T. Buford?

178. Who was "Vinegar Joe"?

179. He had earnestly wanted to lead his victorious men into St. Lô, but was killed after distinguishing himself gallantly just outside the city. When the 29th Infantry Division's victory column moved through the city, it included a lone ambulance bearing a flag-draped coffin. What is its inhabitant called?

180. Who was Iva D'Aquino?

181. Why was "Vinegar Joe" so-called?

182. Who could properly be called "the father of US Army helicopter operations"?

183. Why is John C. Garand famous?

184. Which president helped totally destitute former Confederate officers by allowing

them to receive modest federal government pensions?

185. How did Ernie Pyle die?

186. What Revolutionary War fort was called the "American Gibraltar"?

187. During the Vietnam War, the Army's Special Forces created units called "Mike Forces." What were they?

188. What was a MEDCAP during the Vietnam War?

189. During the Vietnam War, what was the target of the Phoenix program?

190. What was the HES, which was used during the Vietnam War?

191. During the Vietnam War a reporting system called TFES was used. What was it?

PERSONALITIES

1. Identify the author of this famous quotation: "There is nothing so likely to produce peace as to be well prepared to meet an enemy."

2. What was Robert E. Lee's father's name?

3. Although he was personally opposed to slavery and to secession, who felt that he could not honorably fight against the forces of his native state, which had joined the Confederacy?

4. In December of 1953 the Nobel Prize was awarded for the first time to a famous US Army officer. Name him.

5. Name the famous Army general who said: "When things go wrong in your command, start searching for the reason in increasingly larger concentric circles around your own desk."

6. On the occasion of his execution by the British on September 21, 1776, what did Captain Nathan Hale of Connecticut say?

7. What title did Friedrich Wilhelm Baron von Steuben hold in the early Continental Army? What was his chief function?

8. What was the name of the West Point professor who authored the pioneer American study of the art of war?

9. Who was secretary of war during the presidency of Franklin Pierce?

10. Who was the famous general and president who said: "In the final choice a soldier's pack is not so heavy a burden as a prisoner's chains"?

11. What "commanding general of the army" moved his headquarters from Washington DC to New York so that the president and the war department could have less influence over his conduct and policies?

12. Name the decorated Civil War veteran, West Point instructor, tactics innovator and distinguished military scholar who—plagued by the pain of a brain tumor—shot himself in 1884.

13. Who was the first four-star general?

14. Name the famous leader who said: "I cannot trust a man to control others who cannot control himself."

15. Major General Leonard Wood became chief of staff of the Army in 1910. From what college did he graduate?

16. Name the secretary of war responsible for the establishment of the Army War College.

17. Who replaced General Douglas MacArthur as the United States commander in the Far East?

18. Name the famous Army explorers of the Northwest territory.

19. What Revolutionary War figure was nicknamed "Mad Anthony"?

20. Who was traitor Benedict Arnold's accomplice?

21. During what Revolutionary War battle did the famous Molly Pitcher (Hays) replace her mortally wounded canoneer husband?

22. What Army officer is credited with the conquest of yellow fever?

23. Where did Confederate Brigadier General Thomas J. Jackson earn the nickname "Stonewall"?

24. Who said: "In war there is no substitute for victory"?

25. Who was nicknamed "Old Rough and Ready"?

26. In what war did Abraham Lincoln serve?

27. Who is the originator of the famous slogan: "Remember your regiment and follow your officers"?

28. Who was referred to as General Robert E. Lee's "war horse"?

29. What officer and what unit earned the title "Rock of Chickamagua"?

30. Who was promoted to the newly revised rank of lieutenant general and made general in chief of the Union Army in March of 1864?

31. Who said: "No terms except unconditional and immediate surrender can be accepted. I propose to move immediately upon your works"?

32. Who was Clara Barton?

33. Who was the Union general who invented America's favorite pastime: baseball?

34. Who said: "War is hell"?

35. Where did "Jeb" Stuart get his nickname?

36. What Confederate general was nicknamed "Cump" by his family?

37. Who was called the "King of Spades" by his men because of his insistance that they always dig trenches and earthworks?

38. Who resigned from the Army in 1835 to marry the daughter of his commanding officer, Colonel Zachary Taylor?

39. Who was nicknamed "Uncle Sam" by his classmates at West Point?

40. Who was the leader of the Mexican opposition during the US Army's punitive expedition in 1916?

41. Who was Colonel George W. Goethals?

42. Who was the Army's first aviator?

43. Who is responsible for this famous dictum: "There are no bad troops; there are only bad leaders"?

44. Who was General Pershing's aide during the Mexican expedition?

45. In what Army division did the famous Sergeant Alvin York serve?

46. Name the officer court-martialed for his role in the My Lai massacre.

47. What famous WWII general later became secretary of state?

48. Name the most famous WWII cartoonist and artist.

49. Name the speaker and the recipient of this famous order: "Cross the Channel, enter the heartland of Germany and free the continent of Europe"?

50. Name three famous Army generals who later became president of the US.

51. Who captured the "Crown Jewel," Vicksburg, during the Civil War?

52. When ordered to leave the Phillipines by President Roosevelt, General Douglas McArthur made what famous statement?

53. What US Army commander was General Rommel's adversary in the North African campaign in 1942?

54. Who commanded the 5th Army during the assault on Salerno?

55. Who commanded the China-Burma-India Theatre in 1942?

56. Name General Eisenhower's English female chauffeur during WWII.

57. Name WWII's infamous Japanese female propagandist.

58. Who said: "The words 'breaking contact' will no longer be used in this division"?

59. Name the famous leader who said: "Any commander who fails to obtain his objective, and who is not dead or severly wounded, has not done his full duty."

60. Name the Special Forces lieutenant who spent five grueling years as a Viet Cong captive in South Vietnam.

61. Who was Mildred Gillars?

62. What did Robert E. Lee and John McClellan have in common?

63. Name the only five general officers who attained the rank of "general of the armies."

64. How was General George C. Patton killed?

65. Who, during WWII, said: "Look at an infantryman's eyes and you can tell how much war he has seen"?

66. Who said: "In war there is no second prize for runner-up"?

67. Who was the WWI flying ace who later became president of Eastern Airlines?

68. Who said: ". . . There is a limit of human endurance and that limit has long since passed. Without prospect of relief . . . I feel it is my duty to my country and to my gallant troops to end this useless effusion of blood and human sacrifice"?

69. Who commanded the 1st Infantry Division and then later the 104th Infantry Division in WWII?

70. Who was the famous mythical figure of WWII whose name was found in the most unusual places?

71. Name the Korean War figure, hero of the Taejan resistance, POW and Medal of Honor winner.

72. Who was named commander of all UN forces in Korea on July 13, 1950?

73. Who said: "During peace, the Army's primary mission is deterrence—being so well-trained, equipped, and led that no potential adversary would mistake our nation's ability and resolve to defend out interests"?

74 Name the US Army officer who led the first aerial attack on Tokyo in WWII.

75. Who surrendered US forces on Bataan and Manila Bay to the Japanese?

76. For what is William Cummings noted?

77. Name the popular song/slogan created by the heroic defenders of Bataan to identify themselves.

78. What weapon was characteristic of General Matthew B. Ridgeway during the Korean War?

79. What Army officer described his job as "one of the most thankless jobs we can imagine"?

80. Who was Anal Keys?

81. Whose citation by the government of Morocco read: "the lions in their dens tremble on hearing his approach"?

82. Name the first real American war hero in the Pacific in WWII who stayed at the controls of his crashing Flying Fortress to ensure that his entire crew safely escaped.

83. Who was chief of staff of the Army when the attack on Pearl Harbor occurred?

84. Medal of Honor winner Charles E. "Commando" Kelley was cited for his gallantry with the 36th "Texas" Infantry Division during what campaign in WWII?

85. In WWI, Harry S. Truman served with what infantry division?

86. Name the famous organizer of the Rangers who was killed in action while serving as assistant commander of the 10th Mountain Division in WWII?

87. Name the Confederate general who lost a leg at Groveton, Virginia, subsequently used an artificial leg, and had to be strapped to his horse when he mounted it.

88. Who signed the Treaty of Versailles for the US?

89. Name the famous WWI corporal who wiped out an entire German platoon and captured 132 prisoners singlehandedly.

90. Who was Harvey Dunn?

91. Name the author of this 1984 quote: "But the legacy of souls we touch through example setting and through teaching is the long gray line of the Army of the future."

92. Who succeeded General Matthew Ridgeway as commander of the 8th Army in Korea?

93. What was Major Richard I. Bong's claim to fame?

94. Who was Private Paul G. Bennett?

95. Who was President Kennedy's principal military advisor on the situation in South Vietnam?

96. Which general replied "Nuts" when asked to surrender at Bastogne?

97. Who was the commander-in-chief of the German forces in WWI?

98. What nickname did Adolph Hitler give to General Patton?

99. Name the battalion, and its commander, that relieved the 101st Airborne Division at Bastogne.

100. Who created the "Sad Sack" character that represented GI's and their conflicts with the Army?

101. What legendary figure is most closely identified with the entertainment of Army troops in peace and war?

102. Who said: "This . . . is the most honorable place, in the most honorable Army, in the world"?

103. Who said: "There are two kinds of people on this beach! Those who are dead and those who are going to die! Let's move inland!"?

104. For what is Jacob Beser famous?

105. Who succeeded General Westmoreland in 1968 as commander of US forces in South Vietnam?

106. Who was "Cher Ami"?

107. How did Elizabeth Arden, the cosmetic manufacturer, contribute to Army efforts during WWII?

108. Who was the only general officer to land with the first wave on D-Day at Normandy?

109. Who said, "Lafayette, we are here," at the tomb of Lafayette in WWI?

110. Name the famous Japanese general and statesman who favored war with the US and who was executed as a war criminal in 1948.

111. How many "kills" were credited to famous WWI ace Eddie Rickenbacker?

112. Name the famous Civil War heroine who went into battle carrying the colors of her husband's Rhode Island regiment.

113. Who was Private Franklin Thompson?

114. What did Gene Autry do in WWII?

115. Name the commander of US forces in the Dominican Republic.

116. Who was known to his men as "Curly," "Ringlets," or "Old Iron Butt"?

117. Name the last Confederate general to die in the Civil War.

118. Name the Union General who lost an eye, an arm and a foot in three separate Civil War battles before he retired himself from active duty.

119. Name the Army officer who won the Medal of Honor for leading a company-size bayonet charge during the Korean War.

120. Who took over the Far East command from General Ridgeway in May of 1952?

121. Who was Calvin Titus?

122. Who was "Mother Bickerdyke"?

123. Who was Lieutenant General Walter C. Short?

124. Among his many distinctions was the command of the bomber groups that attacked Regensberg and took part in the Tokyo Fire Raid. Who was he?

125. Who predicted in 1924 that one day Japan would go to war against the US and would begin the war with a surprise attack—just after dawn—by carrier-based airplanes against Pearl Harbor and other American facilities in Hawaii?

126. Name the man who was chief of staff of the 42d "Rainbow Division" in WWI.

127. Name the pilot of the **Enola Gay**.

128. Who was Oveta Culp Hobby?

129. The author of **Little Women** and other novels, she also served as a Union nurse during the Civil War. Name her.

130. Known as "Jumpin' Jim," he was described by a colleague, "He could jump higher, shout louder, spit farther, and fight harder than any man I ever saw." Who was he?

131. Complete this statement by General William Westmoreland: "No one has ever demonstrated more ability to hide his installations that the Viet Cong; they were ______________."

132. Name the general who was captured by Communist POW's at Koje-do in May of 1952.

133. Name the general who was detained for five hours by MP's searching for a German posing as a US Army brigadier general during the Battle of the Bulge.

134. Who was Colonel William F. Friedman?

135. Who was the first man to hold the title "Chief of Staff of the Army"?

136. Name the first black general to lead Americans into battle.

137. Who was the first individual to receive the Medal of Honor in the Vietnam War?

138. Name the sergeant of the 165th Infantry Regiment killed on patrol on July 30, 1918.

139. Who was the "Major of St. Lô"?

140. What famous Army officer had the serial number "0-1"?

141. Who said: "No nation was ever blessed with more stout-hearted men than the Americans who fought and died in southeast Asia"?

142. Name the individual who commanded the 1st Battalion, 26th Infantry, 1st Infantry Division in Vietnam and later became Secretary of State.

ARMY LORE

1. When did the first of the M-60 tanks enter the Army inventory?

2. What was a bummer cap?

3. What was a spontoon?

4. What material was used to make canteens until the early 1800s?

5. What was a Brown Bess?

6. What Army entitlement was stopped in 1830 as a result of complaints from officers and a growing social movement?

7. Name three important technological advances that had their first large-scale military application during the Civil War.

8. What is the loaded weight of an M16A1 rifle with a 30-round magazine?

9. Name two major changes made after the first battle of Bull Run.

10. What does the letter "T" on a casualty's forehead mean?

11. What does the acronym SERE stand for?

12. What is an MOS?

13. Once it was called a FEBA. Now it is a FLOT. What is a FLOT?

14. How many articles are contained in the Code of Conduct?

15. What does the phrase "US is always right" mean?

16. What is an ARTEP?

17. Trenchfoot, immersion foot and frostbite are injuries caused by what condition?

18. When was the khaki uniform introduced into the Army?

19. What was a Daugherty wagon?

20. What non-native animal was tested as a pack animal by the Army in Texas in 1856?

21. What is "stand-to"?

22. What are H & I fires?

23. What is an APC?

24. What is the first General Order?

25. What is a company administrative office called?

26. What is the formation of soldiers in preparation for sentry duty called?

27. What is a "fouragere"?

28. A point (man/element) would ________ a unit.

29. What is the maximum effective range of the M-60 machine gun?

30. What do the initials CSMO stand for?

31. How many mils are in a circle?

32. What was a French 75?

33. What is an aiming circle?

34. What bugle call signals the beginning of a formal review?

35. What is a "mustang"?

36. What does a color unit do in a parade or review?

37. Who is sometimes called a "Smokey Bear"?

38. What is an FDC?

39. What is a hip shoot?

40. The infantry color was once white. What is it now?

41. Army vehicles have "cat eyes." What are they used for?

42. What is a guidon?

43. What is a co-ax?

44. In military marching, which foot is used to step-off from a halt?

45. What does the term AWOL mean?

46. What protects the soldier from rain?

47. What does a baseplate support?

48. Who authenticates operations orders?

49. What unique weapon did a horse-cavalry soldier carry?

50. Name the paragraphs in a five-paragraph field order.

51. What is a flack jacket?

52. What is a steel pot?

53. The M203 grenade launcher fires a __________ projectile.

54. Name the term never used to describe a rifle.

55. Name the Army's medium-sized helicopter.

56. What do green felt tabs worn on shoulder loops of Army uniforms signify?

57. Name four types of Army hand grenades.

58. What is "sick call"?

59. True or False: Officers traditionally eat first in an Army mess line.

60. In a marching formation, the company first sergeant would be found at the front or rear of the troops?

61. How are grid coordinates on a military map read?

62. How many artillery pieces would be found in an artillery section?

63. What is a caduceus?

64. The belt tab on an Army belt extends to the wearer's right or left?

65. To what does the term "motor stables" refer?

66. In what war were Browning machine guns and Browning automatic rifles first used?

67. What was a Mill's bomb?

68. What document, usually controlled by the unit first sergeant, announces those assigned to various extra details?

69. A sentry initiates a challenge by saying what?

70. Unit awards are worn on the wearer's right or left?

71. Individual awards are worn on which side of the wearer's uniform?

72. How are bloused trousers worn?

73. An officer's "pinks and greens" uniform was actually composed of what colors?

74. A retreat formation would be held when?

75. A reveille formation would be held when?

76. How frequently was a morning report prepared?

77. What was a T-5?

78. To what does the acronym SPORTS refer?

79. Name three common heat injuries.

80. When a "challenge" is issued, what is returned?

81. On what side of the team did the riders sit on a horse-drawn artillery piece?

82. To what does a "piece" refer?

83. What is a caisson?

84. What is a buck-sergeant?

85. A corporal would have how many stripes?

86. Who is the senior non-commissioned officer in a battalion-sized Army unit?

87. A "top" or "top-soldier" would hold what position?

88. What is an FO?

89. What bugle calls traditionally signal the close of the Army day?

90. An Army second lieutenant would normally command what sized unit?

91. What is a butter-bar?

92. What is a deuce-and-a-half?

93. What does the acronym TOW stand for?

94. What were "conchies"?

95. What is a LAW?

96. What is ZULU time?

97. Once there were eight; now there are three. What are they?

98. Which country invented the tank?

99. Name two utility helicopters.

100. Name the Army's primary attack helicopter.

101. A military compass direction is called a ________.

102. What is the key word used in the collection and reporting of intelligence information?

103. What is the WORM formula?

104. A SADARM is what kind of projectile?

105. What is the Hellfire?

106. What is an FLA?

107. What is a P-38?

108. What is an RPV?

109. What was the nickname of the WWII M-22 light tank?

110. An eight-digit grid coordinate locates a point to the nearest __________ meters.

111. What is determined by the "shadow-tip" field expedient method?

112. What is an FPL?

113. What is the maximum traverse of the tripod-mounted M-60 machine gun?

114. What WWII weapons system was nicknamed the "Calliope"?

115. Name the five major terrain features found on a military map.

116. What WWII vehicle was called the "Aunt Jemima"?

117. Would an individual injured while parachuting from an aircraft that has been brought down by enemy fire be eligible for award of the Purple Heart?

118. What is mask clearance?

119. What does BMNT mean?

120. What is a lightning bug?

121. How many helicopters comprise a light fire team?

122. What name is given to a form of movement in which units are moved successively through one another?

123. What is a T & E mechanism?

124. What does the acronym MOPP stand for?

125. What is an ACE?

126. What is a FIST?

127. Is a deceased member of the Ready Reserves entitled to receive burial honors (pallbearers, firing squad, bugler officer- or NCO-in-charge and chaplain)?

128. What does a properly folded US flag resemble?

129. "Laager" is another term for what?

130. Any locality or area whose seizure or control affords a marked advantage to either opposing force is called what?

131. What is the common name for non-judicial punishment?

132. Name the three types of courts-martial.

133. In the US Army, what is a claymore?

134. How many gunships comprise a heavy fire team?

135. What is enfilade fire?

136. What is the withdrawal of troops by helicopter called?

137. How deep is the fighting position dug?

138. How much time is permitted to properly put on a gas mask?

139. On the day after receipt of notification of death of the president, unless a Sunday or holiday in which case honors will be rendered the following day, the commanding officer of an Army installation will require one gun to be fired at what intervals between reveille and retreat?

140. A cannon salute to the Union consists of how many guns?

141. What was a Black Widow?

142. How many horses are in a cavalry division?

143. What was a DUKW?

144. What was the WWII "Long Tom"?

145. What was called the M-18 "Hellcat" in WWII?

146. What weapon got its name from a gas pipe horn made famous by WWII entertainer Bob Burns?

147. What was the WWII B-17 bomber called?

148. What is the name for an airmobile force on air or ground alert to perform rapid reaction missions?

149. How many US flags are normally permitted to be flown at one time on any Army installation in the US?

150. Name the four types of US flags.

151. What aircraft were called "Grasshoppers"?

152. What WWII vehicle was nicknamed the "Honey"?

153. What weapon did the infantryman of WWI use?

154. What branch of the Army has crossed rifles as its official insignia?

155. What does the number of stars on the shield of an aide-de-camp's insignia mean?

156. A major-general rates a ______-gun salute.

157. What does the term "defilade" mean?

158. What is deadspace?

159. What is beehive ammunition?

160. What was a duster?

161. What was a "potato masher"?

162. Name the weapon first used by Germany in WWI to clear a path through rough, wooded areas held by US forces.

163. Who or what was Big Bertha?

164. What was the foot soldier of WWI affectionately called?

165. What was the B-24 bomber called?

166. What WWII aircraft was called the "Dragon"?

167. What is a slick?

168. What is a dustoff?

169. What does EENT mean?

170. What was a mighty mite?

171. What were "Nike-Ajax" and "Nike-Hercules"?

172. What was Atabrine?

173. How many guns are fired for a national salute to the national flag?

174. Which American WWII bomber was called the "Mitchell"?

175. What is the highest US unit decoration that may be awarded to an Army unit?

176. Who is the only person authorized to award the Medal of Honor?

177. Name the US unit decorations that may be awarded to Army units.

178. What was the B-26 bomber called?

179. What was a Barbette gun?

180. What was a "Rome plow," which was used in the Vietnam War?

181. Are general officers eligible for award of the Combat Infantryman Badge?

182. What was the code name for a B-52 strike in Vietnam?

183. What is concertina wire commonly called?

184. The degree of unit heroism required for award of the Presidential Unit Citation is equal to the degree of individual heroism required for what honor?

185. Name the three degrees of Parachutist Badges.

186. A soldier who fought against the Apaches in 1873 would be authorized to wear what decoration?

187. In a parade formation of regular members of all services, which service marches first?

188. What percent of assigned strength must be awarded the Expert Infantryman Badge for the unit to qualify for the Expert Infantry Streamer?

189. A unit patch worn on the right shoulder sleeve means service with that unit in __________?

190. What is the "President's Hundred Tab"?

191. In a parade formation in which cadets, midshipman of all services and regular members of all services are participating, which element would be first in the line of march?

192. Originally designed as an anti-aircraft weapon, it had twin 40-mm cannon mounted on a tank chassis. What was it called?

193. What is a recon sergeant?

194. What is a TOC?

195. What is an "aiguillette"?

196. What is a PLL?

197. What is an ASL?

198. What is a TAOR?

199. What is an AO?

200. What do the initials LRRP stand for?

201. What was "Puff the Magic Dragon"?

202. Traditionally, a newly commissioned officer gives a dollar to the first man who __________ him?

203. What are pathfinders?

204. What is nap-of-the-earth flight?

205. During airmobile operations, what would a single trip from one area to another, regardless of the number of helicopters involved, be called?

206. What is a RAOC?

207. What series of vehicles is produced for the Army by the FMC Corporation?

PHOTOGRAPHS

1. Who is the famous general shown?

2. How old was he at the time this photograph was taken?

3. To what unit does the pictured guard belong?

4. What is his unit's nickname?

5. Who is the famous general shown? 6. To what unit do these troops belong?

7. What is the name of the Army post shown?

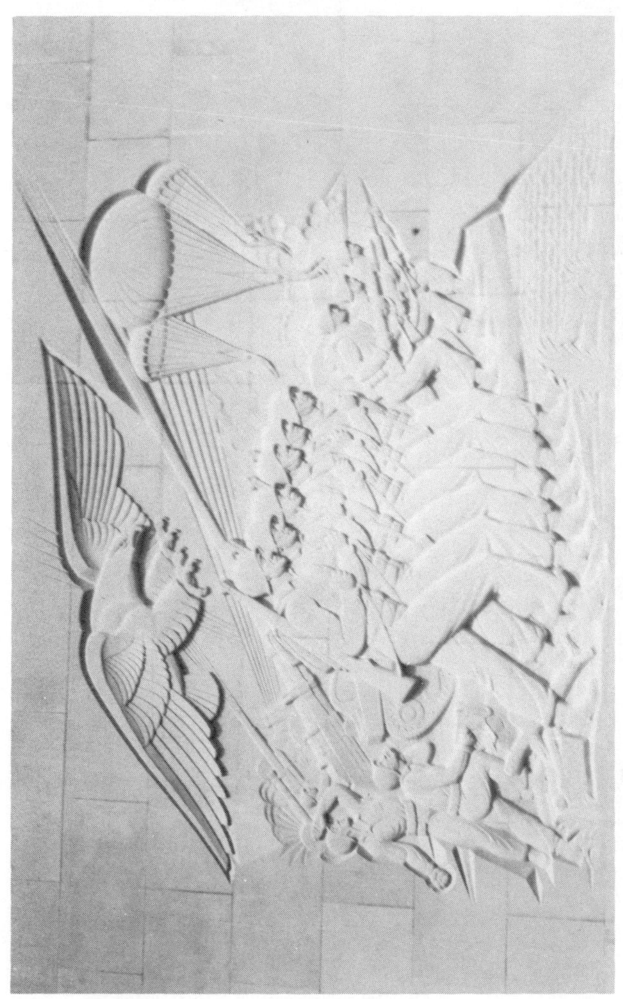

8. Where is this sculptured relief located?

9. What is the name of the general in the photograph?

10. Where is Messina?

11. Identify this photograph. 12. Where was this photograph taken?

13. Who is speaking? 14. To whom is he speaking?

15. Where is this building located?

16. Name the three generals in the Jeep.

17. Where is this building located?

18. Name the two generals in the photograph.

19. Where was this picture taken?

20. Which medal is the Army Medal of Honor?

21. When was the Army Medal of Honor authorized?

22. Who is the American general shown?

23. Identify this building.

24. Where was this photograph taken?

25. Where was this photograph taken?

26. What is the name of this statue? 27. Where is it located?

28. What is the name of this monument?

29. Where is it located?

30. Where is this building located?

31. Where was General Dwight D. Eisenhower born?

32. What was his rank upon retirement?

33. What Army post is pictured?

34. What was the former name of this building?

35. Where is it located?

36. Where was this photograph taken?

37. What monument is pictured?

38. Where is it located?

39. Where is this building located?

40. What unit is stationed at this post?

41. What is the name of this residence?

42. Where is it located?

43. Where is this building located? 44. Which building is it?

45. What is the name of this statue?

46. Where is it located?

47. Who is this man?

48. From what college did he graduate?

49. What is the tank pulling in this photograph?

50. In what war was it used?

51. What is the name of this statue?

52. Where is it located?

ARMY HISTORY

1. Name the official Army song.
2. What is the birthday of the United States Army?
3. Name the two oldest units in the US Army.
4. What event occurred on St. Patrick's Day (March 17th) in 1776?
5. What was the title of George Washington's treatise on the defense of America?
6. Name the oldest organization in the Regular Army.
7. Who is considered to be the father of the US Military Academy at West Point?
8. What was the "Grand Army of the Republic"?
9. Name the Army's newest branch.

10. Name the first American unit sent to France in WWI.

11. What were "The Immunes"?

12. By what name is the Militia Act of January 21, 1903 commonly known?

13. Name the law passed in August of 1912 which governs service on the Army General staff.

14. In 1926 Congress changed the name of the Air Service to what?

15. What agency was created on August 10, 1949?

16. By what generic name were the earliest American military organizations called?

17. What are the words on the official seal of the Army?

18. How many battle streamers are displayed on the US Army flag?

19. What event occurred three weeks prior to the American victory at New Orleans in the War of 1812?

20. Who founded the Army Nurse Corps?

21. What distinction applies uniquely to Dr. Mary Walker?

22. The Treaty of Guadelupe Hidalgo ended what war?

23. How long did the Mexican siege of the Alamo last?

24. What did an Executive Order signed by President Eisenhower promulgate?

25. Name the organization—higher than a corps—created by Major General Ambrose E. Burnside to achieve greater ease of tactical control.

26. What Army regiment participated in the successful attack on Tientsin during the Boxer Rebellion?

27. What legislation provided the National Guard with Federal funds, prescribed twice-a-month drills and annual training, and patterned the National Guard's organization and equipment after that of the Regular Army?

28. What Army branch began as a section of the Signal Corps?

29. How long did it take the Army to build the Panama Canal?

30. What occurred on the nights of July 9 and 10, 1943?

31. What organization built the north and south wings of the US Capitol in Washington DC?

32. On what date was the Army Flag dedicated?

33. Name the flag hoisted by General Washington in January of 1776 at

Cambridge, Massachusetts as the standard of the Continental Army.

34. What is DA?

35. What unit was "First at Vicksburg"?

36. Name the first US Army unit committed to offensive combat in WWI.

37. To what groups does the term "Total Army" refer?

38. Until August of 1918, there were three categories of Army divisions in WWI. Name them.

39. When did Secretary of Defense McNamara propose a complete reorganization of the US Army Reserve and Army National Guard?

40. When was the Uniform Code of Military Justice enacted?

41. When did US Army forces leave Iceland?

42. What did the Hoover Committee recommend renaming the National Military Establishment?

43. The WWI Selective Service Act (draft) expired in what year?

44. In what year did the US Army end its occupation of Germany after WWI?

45. Germany requested an armistice in 1918 on the basis of what document?

46. When was West Point founded?

47. What was the title of the senior officer of the Army from 1775 to 1903?

48. Name the city where Roosevelt, Churchill and the combined chiefs of staff met to plan future WWII strategy.

49. Name the emergency legislation that gave the US president the authority to provide goods and services to other nations considered vital to the defense of the US.

50. Name the plane that dropped the first atomic bomb.

51. The outbreak of the Korean War caused what to be extended until 1959?

52. In what year was an Army draft lottery adopted?

53. When was the Women's Army Corps established?

54. Which Army Reserve division was operating an Army training center three weeks after being mobilized during the Berlin build-up?

55. Name the Army general who was present and received one of the five fountain pens General MacArthur used to sign the instrument of surrender with the Japanese.

56. In 1890 what kind of cloth was formally adopted by the Army for uniforms?

57. What date is the official birthday of the United States Army Reserve?

58. Who was the author of the "Expansible Army" plan?

59. Name the three states that opposed the call for the states to raise manpower for the War of 1812, maintaining that the call was unconstitutional and illegal.

60. Name the Civil War Union unit whose 864 members each received the Medal of Honor.

61. What was the common name given to the 1st Regiment of the United States Volunteer Cavalry?

62. What was distinctive about WWII's 99th Pursuit Squadron, which fought in Italy?

63. Who was the highest ranking Army officer to be killed in WWII?

64. What was the password used by the Army on D-Day, June 6, 1944?

65. What was the Cuban Expeditionary Force?

66. Name the three full generals of WWI.

67. In what year was the legislation passed that renamed the various reserve corps the "Army Reserve"?

68. For what action was the first Army battle streamer awarded?

69. How many battle streamers were added to the Army flag during the Revolutionary War?

70. How many battle streamers were added to the Army flag as a result of the Boxer Rebellion in China?

71. On August 16, 1903 the first ________ ____________ took office.

72. How many battle streamers on the Army Flag are for the Spanish-American War?

73. What was abolished in the Army on August 5, 1862?

74. When was Morse code visual signaling first used by the Army?

75. In what year was "blue" specified as the national color for Army uniforms?

76. What event occurred on August 24, 1814?

77. When was the Chaplain Corps established?

78. In what year did the Judge-Advocate General's Department become a corps?

79. The Intelligence and Security branch of the Army became the Military Intelligence branch in what year?

80. What was the Corps d'Afrique?

81. When did the use of serial numbers for enlisted men begin?

82. What information was contained on the first dog tags used by the Army?

83. True or False: When initially authorized, the Medal of Honor could only be awarded to NCOs and privates.

84. When General Eisenhower had his tailor modify his bulky officer's jacket, what did he create?

85. When was the Field Artillery branch established?

86. Did the Battle of the Little Big Horn result in a battle streamer for the army flag?

87. When was the MP Corps established?

88. The Corps of Engineers dates from ____________?

89. How many battle streamers were added to the Army flag for the Indian Wars?

90. What colors are the Army's battle streamers from the Civil War?

91. How many battle streamers were added to the Army flag after the Civil War?

92. Who was Major General William C. Lee?

93. As commander of a tank unit in Africa, who was the first US senator to see combat action since the Civil War?

94. Name the two infantry regiments that participated in the Siberian Expedition in 1918.

95. How many battle streamers were added to the Army flag for campaigns in the Mexican War?

96. What was the "Society of Harmonious Fists"?

97. When was the Chief of Staff of the Army authorized to be a general (four stars)?

98. When were divisional shoulder patches authorized for the Army?

99. Who built the Panama Canal?

100. What color was adopted for uniforms by the Army in 1902?

101. Name the Army private executed in 1945 for desertion.

102. What is the name given to the slaughter of 101 American prisoners by German SS troops in Belgium in late 1944?

103. How many battle streamers were added to the Army flag after the War of 1812?

104. How many battle streamers were added to the Army flag for the Phillipine Insurrection?

105. How many battle streamers were added to the Army flag for WWI?

106. In 1942, Stalin (with tongue in cheek) agreed that after WWII Korea should be

an independent republic. Where was this conference held?

107. On August 2, 1945, where did the US, Russia and England hold a conference to effect the agreements previously reached at Yalta?

108. How many battle streamers were added to the Army flag for WWII action?

109. How many battle streamers were added to the Army flag for the Korean War?

110. How many divisions did the Army employ in the Korean War?

111. Some view the _______________ as one of MacArthur's major tactical errors in the Korean War.

112. Any US Army member who served at the Front in WWI was automatically entitled to belong to what famous organization?

113. What is May 12, 1945 called?

114. Name the US Army transport ship on which the heroic "Four Chaplains" gave up their life jackets so that others might live.

115. What event occurred on June 25, 1950?

116. What division is credited with the capture of the Remagen Bridge?

117. The WWI Armistice Treaty was signed in a railcar located in what forest?

118. How many Allied powers occupied Germany after WWII?

119. When the US Army flag is displayed in a stationary position, the streamer embroidered with what words should be in the center facing foward and completely identifiable?

120. Were any female Army personnel killed by hostile fire in Vietnam?

121. When did the Air Force separate from the Army?

122. Name the only two WWII unnumbered divisions.

123. During what war did the first use of unit patches occur?

124. In what year were the "Geneva Accords" ending the Indochinese War between France and Vietnam signed?

125. Where did Roosevelt, Churchill and Chiang Kai Shek meet to pledge continuation of the war against Japan until unconditional surrender?

126. Before they were called brigades, what were the major subordinate elements of the Armored Division called?

127. How long was the initial leg of the infamous "Bataan Death March"?

128. What was the WWII "Devil's Brigade"?

129. What social reform in the Army was precipitated by the Korean War?

130. What was different about the 99th Infantry Battalion in WWII?

131. How were the six regiments of Confederate prisoners utilized in the Civil War?

132. Give the effective date and time of the WWI armistice.

133. What was Theodore Roosevelt referring to when he called it ". . . one of the most iniquitous bits of legislation ever placed on the statute books"?

134. When was the Infantry branch established?

135. What distinguished the Army's 92d and 93d Infantry Divisions in WWII?

136. What was the coalition of Germany, Italy and Japan, which lasted from 1936-1945, called?

137. How many battle streamers were added to the Army flag after the Grenada rescue operation?

138. How many battle streamers were added to the Army flag for the Vietnam War?

139. Who raised the battleship **Maine** from the bottom of Havana Harbor in 1911?

140. What was the name of the daily newspaper for US Armed Forces in WWII?

141. When did the Grenada operation commence?

142. Of the eight major wars since the Revolution, five were fought under a Congressional Declaration of War. Name the three which were not.

143. What was the only occasion in US history where reserve forces were mobilized as a pure instrument of foreign policy?

144. What unit formed the Senior Army Headquarters during the Dominican Crisis of 1965-66?

145. What was the initial mission of Army forces in the Dominican Crisis of 1965?

146. Name the countries which participated in the OAS peacekeeping force in the Dominican Republic.

147. What allied country provided the first troops in support of US Army efforts in South Vietnam?

148. How many reserve component divisions were mobilized for the Berlin build-up in 1961?

149. How many National Guard divisions were mobilized during the Korean War?

150. Name the group which recommended the addition of large numbers of helicopters to Army divisions.

151. How many combat arms regiments are now contained in the new US Army Regimental System?

152. To what does the term CARS refer?

153. Name the joint project of the US Postal Service and the Army Air Corps that was inaugurated in May, 1918.

154. What was Operation Downfall?

155. During WWII, a small flag with one or more blue stars on it placed in the front window represented __________ serving their country.

156. During WWII, a small flag with one or more gold stars on it placed in the front window represented ________ serving their country.

GENERAL

1. Who composed "The Caisson Song"?

2. Where is the US Army Chaplain School located?

3. What insurrection occured in western Pennsylvania during the summer of 1794?

4. What is the present name of the School for the Regiment of Artillerists and Engineers, established in 1794?

5. What practice, adopted by the Regular Army in 1806 and modeled after the British system, awarded honorary rank to an officer who was cited for meritorious conduct?

6. US Army personnel attending the Defense Language Institute would go to school in either of what two places?

7. Where is the US Army Strategic Studies Institute?

8. Name the two practices, inherited from the militia system, that were finally stopped during the Civil War.

9. What famous Army institution was established by presidential order in 1895?

10. Name the "Marne Division."

11. Who founded what is now called the Command and General Staff College at Fort Leavenworth, Kansas?

12. Why is a lieutenant-general senior to a major-general?

13. Where is the US Army Signal School located?

14. What was the name of George Armstrong Custer's regiment?

15. How many divisions did the Army mobilize in WWII?

16. In WWI the Army fought with a "square" division. What type of division was used in WWII?

17. Where is the US Army Ordnance School located?

18. For what do the initials AEF stand?

19. What is the Badge of Military Merit, established by General George Washington on August 7, 1782, now called?

20. What famous Indian chief, leading a confederation of American and Canadian Indians as British allies, was killed at the Battle of the Thames in the War of 1812?

21. What is the Continental Cannon Foundry located in Springfield, Massachusetts now called?

22. Where is the US Army Armor School located?

23. Name the "Ivy Division."

24. During the Peninsular Campaign of the Civil War, Brigadier General Daniel C. Butterfield devised a now famous bugle call. Name it.

25. What was authorized by Congress in 1862 and first awarded in 1863?

26. What was the name of Robert E. Lee's horse?

27. What were "Blue Cross," "Green Cross" and "Yellow Cross"?

28. What was the "TR Division"?

29. What was sag paste?

30. Where are Army medical schools located?

31. Name the "All American Division."

32. What was the Freedmen's Bureau?

33. What barrier is called the “Z” by GI’s?

34. What division drove into Belgium in WWII to relieve the defenders of Bastogne?

35. Where is the US Army Chemical School located?

36. An adjutant was once a principal assistant; now he is charged with what two functions?

37. Where is the US Army Ranger School located?

38. “Garry Owen” is the official song of what unit?

39. Federal law prohibits the overseas deployment of a soldier who has less than how many days of training?

40. What is frocking?

41. In WWI, the 2d Division was composed of Army and what other troops?

42. What is called the “King of Battle”?

43. What day is celebrated by Army units with open-houses, contests, picnics, etc.?

44. What musical interlude is played during ceremonies where a general officer is present?

45. How many volleys of rifle fire are there at a military funeral?

46. The "Tropic Lightning Division" is what infantry division?

47. What rank of officer commands a division?

48. Who is the patron saint of the artillery?

49. The beak on a colonel's insignia of rank faces in which direction?

50. Name the "Bayonet Division."

51. What is the armor and cavalry color?

52. Where is the US Army Artillery School?

53. What term is used to describe military and civilian preparations for war?

54. For what does NORAD stand?

55. What size unit would a lieutenant general command?

56. Name four types of Army divisions.

57. What is the Signal Corps' color?

58. What legislation initiated what has become the Reserve Officers Training Corps?

59. There are three reasons for award of the Bronze Star Medal. Name them.

60. What US Army school is co-located with the University of Virginia?

61. "In the everlasting glory of the Infantry, stands the name of . . ." — finish this quotation.

62. What US Army school is located at Fort Eustis, Virginia?

63. With what division is the word "nuts" associated?

64. What two words are needed to complete this sentence: "I am the Infantry, ______________!"?

65. What unit is nicknamed the "Rakkasans"?

66. What is the highest enlisted US Army rank?

67. An Army lieutenant colonel would command what size unit?

68. Who is the "Queen of Battle"?

69. What rank of officer wears silver oak leaves?

70. The US Army has commissioned officers and what other two kinds of officers?

71. An Army captain would command what size unit?

72. What was a "sutler"?

73. What was the Army's peak strength during the Vietnam War?

74. Which regiment is called the "Blue Spaders"?

75. Name the "Victory Division."

76. Name four types of infantry units.

77. Correlate the rank with the number of stars.

1. Lieutenant General	a. Five
2. Major General	b. One
3. General of the Armies	c. Three
4. General	d. Two
5. Brigadier General	e. Four

78. Name the "Rainbow Division."

79. Name the famous unit commanded by General Claire Chennault in WWII.

80. An Army colonel would command what size unit?

81. Where is the Army Air Defense School now located?

82. Where is the Army Military Police School located?

83. The "Black Widow" WWII fighter was a __________.

84. Where is the "Home of the Infantry"?

85. What is located at Fort Rucker, Alabama?

86. What Army school is now located at Carlisle Barracks, Pennsylvania?

87. When was horse cavalry abolished?

88. Name the "Hell on Wheels Division."

89. What was the peak amount of US Army manpower during WWI?

90. Name the "First Team Division."

91. Where and when did the First Division earn the nickname "The Big Red One"?

92. Where is the US Army Quartermaster School?

93. Who wrote the "Ballad of the Green Berets"?

94. Where is the US Army Finance School?

95. Applicants to West Point must be between what ages?

96. Name the "Spearhead Division."

97. What school is located at Fort Huachuca, Arizona?

98. What year was the Army's Canine Corps established?

99. Name the "Red Diamond Division."

100. Where is the US Army Special Forces School?

101. What was the peak US Army strength during WWII?

102. Where is the US Army Command and General Staff College located?

103. Name the "Indianhead Division."

104. What did the Rivers and Harbors Act of 1824 do?

105. What WWII division was called the "Red Bull Division"?

106. Name the famous "Texas Division."

107. How many appointments to West Point are available to sons or daughters of persons awarded the Medal of Honor?

108. What was the 37th Infantry Division called?

109. Which WWII division was called the "Cyclone Division"?

110. A West Point graduate must serve a minimum of how many years with Army?

111. What was "Cochran's Convent"?

112. What is the nickname of the 40th Infantry Division?

113. A famous WWI poster promised "________________" for the man who served his country?

114. What famous American artist designed the music coversheet for the popular WWI song "Over There"?

115. What is the popular name of the US Armed Forces Servicemen's Readjustment Act of 1944?

116. A glider patch is worn on the overseas cap by ____________ qualified personnel under certain conditions.

117. What unit is called the "Blue and Gray Division"?

118. What was the nickname of the 45th Infantry Division?

119. What was the 31st Infantry Division called?

120. Name the tavern in New York City where George Washington "took leave of his officers."

121. What WWII division was called the "Blood and Fire Division"?

122. The Vietnam War continued through how many US presidential administrations?

123. Which division is called the "Old Hickory Division"?

124. "The First Team" is the nickname of what division?

125. How many stars are on the blue silk shield from which the Army Medal of Honor is suspended?

126. Name the second oldest Army branch.

127. What was disestablished on October 20, 1978?

128. To what Mexican War unit was Zachary Scott referring when he said, "Brave Rifles! You have been baptised in blood and are brought forth steel!"?

129. In what year did the first female officers graduate from West Point?

130. What is the name of the Army's Parachute Team?

131. What is the nickname of the 1st Armored Division?

132. What is the Society of Cincinnati?

133. What is the word inscribed on the suspension bar of the Army Medal of Honor?

134. Which WWII division was called "The Angels"?

135. To whom did the name "Reconstruction Aides" refer?

136. Were woman enlisted in the Army during WWI?

137. Name the "Liberty Bell Division?"

138. Who was represented on the Army WAC insignia?

139. What were WASP s?

140. Which division was called the "Trailblazer Division"?

141. What is the nickname of the 3d Armored Division?

142. What division is called the "Breakthrough Division"?

143. What WWII division was called the "Statue of Liberty Division"?

144. How many infantry divisions were deployed in France during WWI?

145. What division was called the "Lightning Division"?

146. How many infantry divisions were organized for service in WWI?

147. What was the nickname of WWII's 79th Infantry Division?

148. What is significant about the towns of Columbus, New Mexico and Glen Springs, Texas?

149. What was the Anaconda Plan?

150. The Chemical Warfare Service was organized on June 28, 1918. It was redesignated as the Chemical Corps in what year?

151. From 1917 to 1957, the US Army Signal Corps maintained a unique system of message carriers. Name it.

152. What Army branch was established on June 21, 1860?

153. What was the 81st Infantry Division of WWII called?

154. When did the first use of "mines" occur?

155. During the Vietnam War, what was the name given to Army volunteers who searched the enemy's underground complexes?

156. What was a pigeon nest?

157. What is the common name for the UH-1 Iriquois helicopter?

158. How many training divisions are in the United States Army Reserve?

159. What was significant about San Isidro Airfield?

160. What was the nickname of the 80th Infantry Division?

161. What is the name of the Air Force pilots who direct the fires of close air support aircraft for the Army?

162. What was a pigeoneer?

163. Name the two National Guard divisions mobilized for the Berlin build-up?

164. What does REFORGER stand for?

165. What organization was established on May 14, 1942 as an auxiliary of the Army?

166. In 1776 it was called cavalry. In 1918 it became the tank corps. In 1940 its name was changed to the armored force. What branch is it now?

167. What was WWII's 83d Infantry Division nicknamed?

168. In what year was legislation passed permitting females to attend West Point?

169. What was the "Legion of the United States"?

170. What division is credited with being the first to wear a shoulder patch?

171. What is the "Yankee Division"?

172. Name the organization that became a component of the Army on July 14, 1943.

173. What WWII division was called "The Railsplitters"?

174. Name the organization of former WASP's.

175. What was the nickname of the 86th Infantry Division?

176. Why is the West Point dress uniform gray?

177. What branch of the Army has a benzene ring on its insignia?

178. Name the "Cloverleaf Division."

179. In the days when the Army had regiments, why was there no "J" company?

180. With which infantry division is the slogan "These are my credentials" associated?

181. What was the payload of a WWII CG-4A combat glider?

182. What is "Three Cheers"?

183. Which WWII division was called the "Hellcats"?

184. What is the phrase "by Direction of the President (DP)" used for in the Army?

185. What were "Comrades in Battle"?

186. What branch of the Army has the Harper's Ferry pistol on its insignia?

187. When were campaign medals authorized for the Army?

188. What division was called the "Rolling W Division"?

189. When were the first female cadets admitted to West Point?

190. What was a ruptured duck?

191. What was the 90th Infantry Division nicknamed?

192. What was a "Donut Dolly"?

193. Air Defense Artillery separated from Field Artillery and became a separate branch of the Army in what year?

194. What is the name of the Command and General Staff College yearbook?

195. The "Buffalo Division" of WWII was which infantry division?

196. What was the "Chieu Hoi" program?

197. To what did the term CMS, used during the Korean War, refer?

198. Which WWII division was called the "Powder River Division"?

199. Who wrote the popular WWII song, "This Is Your Army, Mr. Jones"?

200. What unit was nicknamed the "Victory Division"?

201. All units of the US Army are designated by number except what size units?

202. The "Deadeye Division" was which infantry division?

203. Founded on February 2, 1901, it was the first women's component of the Army. Name it.

204. Name the infantry division that cleaned up Bura-Gona beach head and reopened the Phillipines for General McArthur.

205. What WWII division was called the "Iriquois Division"?

206. How many West Point appointments are authorized to be made by US senators?

207. The "Trident Division" was which infantry division?

208. What branch insignia contains the symbol of Helios, Greek god of the sun?

209. Where is the US Military Academy Preparatory School located?

210. From 1954 to 1974 the National Guard constituted over fifty percent of what branch of the Army?

211. The numerical designation of a corp is written in ____________ numerals.

212. Who/what is often described as "the ultimate weapon"?

213. What was the PROVN Study?

214. What was SOG?

215. How many Army reserve component units were mobilized and deployed during the Vietnam War?

216. Where was Fury Drop Zone?

217. Which WWII division was nicknamed the "Century Division"?

218. Name the Confederate prison where most Union officer POW's were held.

219. The "Ozark Division" was which infantry division?

220. How many appointments to West Point are authorized to be made by the president?

221. Name the four degrees of the Legion of Merit.

222. What is the Medal of Honor Roll?

223. What was "klim"?

224. What was the "Ruptured Duck"?

225. What was "Pick's Pike"?

226. What was the nickname given to the board of officers convened after the Civil War to reduce the size of the officer corps?

227. What was marston mat?

228. What is a "sky pilot"?

229. What was the P-38 fighter called?

230. What was the name of Army organization, born during the Phillipine Insurrection, which was initially officered by Americans but was composed of native soldiers?

231. What was "Yellow Jack"?

232. The beak on "War Eagles" faces in which direction?

233. In March, 1953 at Panmunjon, 6,670 sick and disabled Communist prisoners were traded for 684 Allied prisoners in an operation called __________.

234. What was General George S. Patton's white English bull terrier named?

235. Who was Rita Louise Zucca?

236. Who was Colonel Phillip C. Cochran?

237. Who was "Veronika Dankeschön?

238. **See Here,** ____________________ was a best-selling WWII book about life as a draftee in an Army training camp.

239. The WWII P-51 fighter was also called a __________.

240. What was "Operation Big Switch"?

241. What was the 104th Infantry Division called?

242. Name the organization founded in 1879 to provide Army officers with insurance when commercial companies either refused to insure them, or charged them exorbitant rates.

243. Which automaker built the first Jeep?

244. Was "Buffalo Bill" Cody awarded the Army Medal of Honor?

245. What is the Hughes Trophy Award?

246. A "liver patch" is the irreverent nickname for what device?

247. Whose profile is on the Army Medal of Honor?

248. Name the "Old Ironsides Division."

ANSWERS

CONFLICT

1. It was the last major battle of the Civil War.

2. April 19, 1775

3. Lexington, Massachusetts

4. Breed's Hill in Charlestown, Massachusetts

5. The Battle of Fort Moultrie

6. George Washington's crossing of the Delaware River

7. Valley Forge, Pennsylvania

8. Yorktown, Virginia

9. The battle of Chippewa on July 5, 1814

10. The Seminole War

11. The Mexican War (1846-1848)

12. The Battle of Wounded Knee

13. The "Mormon War"

14. The Spanish-American War

15. The Americans

16. A bald eagle—the mascot of the Union's "Iron Brigade (Eagle Brigade)" in the Civil War. From his perch atop a long pole, Old Abe and his screeching inspired the brigade through thirty-six battles. He was wounded twice.

17. The surrender of Charleston on May 12, 1780

18. The massacre of wounded Kentucky militia at Frenchtown, Michigan by Indian elements of British forces during the War of 1812

19. Union forces tunneled under Confederate lines and detonated 7,000 pounds of gunpowder to breach confederate defensive works.

20. Antietam

21. Pickett's Charge at Gettysburg

22. After the fall of Fort Sumter, in which there were no casualties, a Confederate private was killed during ceremonial salutes fired as the fort was occupied by the South.

23. The Forty-third Partisan Rangers

24. Bragg (Captain Braxton Bragg)

25. Comanche

26. Lieutenant colonel
27. Apache
28. Brigadier General John J. Pershing
29. The French and Indian War
30. The Korean War
31. Gas
32. Battery C, 6th Field Artillery
33. The Battle of Corinth
34. General Joshua L. Chamberlain, Colonel of the 20th Maine Volunteers, who received the Medal of Honor for his defense of Little Round Top, was wounded six times, and was promoted to Brevet Major General. After the Civil War, he was elected Governor of Maine four times and became president of Bowdoin College.
35. France and Spain
36. A reserve force of 10,000 militia from Pennsylvania, Maryland and Delaware
37. Shawnee
38. General Ambrose Burnside
39. General William Sherman
40. Fifth Corps
41. Clark Air Field, the Philippines

42. The Civil War

43. El Alamein

44. Vichy

45. The Maginot Line

46. Hirohito

47. "Blitzkrieg"

48. General Petain

49. Guadalcanal

50. Inchon

51. KATUSA'S

52. T-34

53. Pusan

54. Subterranean

55. Heartbreak Ridge

56. Atomic war

57. Combat parachute jumps by the 187th Airborne Regimental Combat Team

58. General MacArthur's first amphibious plan in the Korean War. It was to have involved the 1st Cavalry Division and a Marine regimental combat team, but had to be scrubbed due to the necessity of committing the 1st Cavalry to the Pusan Perimeter.

59. Task Force Smith—elements of the 1st Battalion, 21st Infantry, 24th Infantry Division, with mortar, recoilless rifle and bazooka reinforcements, commanded by Lieutenant Colonel Charles B. Smith

60. The 7th Infantry Division

61. Iwon

62. The 24th Infantry Division

63. September 15, 1950

64. The 45th Infantry Division

65. The Meuse-Argonne offensive of WWI, where more than 1,200,000 Americans advanced on a 90-mile front

66. 187

67. 120,000

68. Civilian Irregular Defense Group

69. The Third Infantry Division

70. A task force of the 3d Infantry Division, which fought inland to assist X Corps during its evacuation to Hungnam

71. A line of fire—bases, sensors, radars and patrols—below the DMZ designed to prevent enemy infiltration into South Vietnam

72. The 101st Airborne Division

73. Regional Forces (RF) and Popular Forces (PF), which were South Vietnamese local militia

74. Dak To

75. Junction City

76. The Fifth

77. The Battle of the Bulge

78. 1st Infantry, 4th Infantry, 9th Infantry, 25th Infantry, Americal, 1st Cavalry (Airmobile), 101st Airborne

79. 4,000

80. On the banks of the Little Big Horn River in Montana

81. The 38th Parallel

82. 173d Airborne Brigade; 3d Brigade, 82d Airborne Division; 1st Brigade, 5th Infantry Division (Mech); 199th Light Infantry Brigade; 198th Light Infantry Brigade; 196th Light Infantry Brigade; 11th Light Infantry Brigade; 11th Armored Cavalry Regiment

83. Russian

84. The 7th Infantry Division

85. Hungnam

86. Chipyong-ni

87. The drive by South Vietnamese forces with US Army advisors and support into Laos to cut the Ho Chi Minh Trail

88. The Son Tay Raid

89. They were the names of the areas in Cambodia that were invaded on May 1, 1970.

90. An enormous enemy basecamp near Snoul which contained more than 300 bunkers, 500 huts and miles of tunnels and trenches

91. The Tet Offensive

92. "Riverine Operations"

93. "Over There"

94. The bayonet

95. 180,000 men

96. August 6, 1945

97. A "Human Sea"

98. "Position Warfare"

99. 16,000

100. The 17th parallel

101. "The Desert Gallop"

102. At Remagen

103. At Eisenhower's headquarters in the red schoolhouse at Reims, France

104. A "balloon bomb," which exploded in Oregon on May 5, 1945

105. The Ludendorff Bridge

106. The Siegfried Line

107. They were inland from Normandy participating in war games.

108. "The Longest Day"

109. Sword, Juno, Gold, Utah and Omaha

110. It was the Japanese POW camp that held the Army nurses captured on Bataan and Corregidor.

111. Patton's Third Army

112. The Vietnam War

113. 11th Airborne Division

114. Okinawa

115. Four

116. "Iron Triangle"

117. "Thunder Road"

118. The 1st and 2d Battalions, 75th Rangers

119. Six months

120. At Kasserine Pass

121. Germany's network of defensive fortifications, almost four miles deep, stretching across the Italian penninsula

122. Okinawa

123. To the Dominican Republic

124. "DMZ" (Demilitarized Zone)

125. The Ia Drang Valley

126. The 2d Battalion, 503d Infantry, 173d Airborne Brigade

127. Ben Suc

128. "International Safety Zone" (ISZ)

129. Utah and Omaha

130. The 2d Division

131. WWI

132. It is a strategically important prominence near but separate from San Juan Hill. It was the actual objective that was attacked by the 9th Cavalry, part of the 10th Cavalry and by Lieutenant Colonel Theodore Roosevelt's "Rough Riders" during the Spanish American War.

133. The Meuse-Argonne Offensive

134. The Saint-Mihiel Offensive

135. The Allies

136. The "Red Ball Express"

137. The 3d Division

138. The house of Wilmur McLean in the town of Appomattox Court House, Virginia

139. The Battle of Vicksburg

140. Seventeen

141. 303

142. "Climb Mount Niitake"

143. "Merrill's Marauders"

144. "Banzai"

145. Seventy-two

146. Sicily

147. The Arc de Triomphe

148. "Meat Grinder"

149. An WWI organization of volunteer US aviators who flew for the Allies prior to the US declaration of war

150. "Hat-in-the-Ring-Pursuit Squadron"

151. On the Swiss border

152. Koje Island

153. 11,500

154. Retractable bridges

155. "Operation Strangle"

156. Seven

157. The 36th "Texas" Infantry Division

158. Several units are referred to in this way, among them the 2d Battalion of the 19th Infantry and elements of the 1st Battalion of the 117th Infantry.

159. A WWII Army Air Force unit whose mission was to airdrop arms and supplies to resistance fighters behind enemy lines

160. The formal title of Merrill's Marauders

161. "Brimstone"

162. "Destiny"

163. General Dwight D. Eisenhower

164. The 442d Regimental Combat Team, composed of Nisei-Americans

165. Ledo Road

166. Kasserine Pass

167. The Abbey of Monte Cassino

168. The Himalayas

169. Hürtgen Forest

170. Betty Grable

171. The Falaise Gap or Pocket

172. The "Malinta" Tunnel

173. Anzio

174. Leyte Island

175. The "Bloody Bucket"

176. Fort Ticonderoga

177. The alias used by Loreta Janeta Velasquez, a spy for the Confederate Army in the Civil War

178. General Joseph Stillwell

179. "The Major of St. Lô"

180. Tokyo Rose

181. He had a sharp tongue and a sharp retort.

182. Major General Harry W.O. Kinnard

183. He invented the M-1 (Garand) rifle.

184. Ulysses S. Grant

185. He was killed by a Japanese sniper on the island of Ie Shima on April 18, 1945.

186. Fort Ticonderoga

187. Mobile Strike Forces, quick reaction forces of indigenous personnel designed to quickly relieve or reinforce a threatened Special Forces camp

188. Medical Civic Action Program

189. The Viet Cong Infrastructure (VCI)

190. The Hamlet Evaluation System, an automated procedure used to calculate the degree of pacification of South Vietnam

191. The Territorial Forces Evaluation System, a device used to measure the effectiveness of the RF/PF forces

PERSONALITIES

1. General George Washington
2. Henry "Lighthorse Harry" Lee, a general in the American Revolution
3. Robert E. Lee
4. George C. Marshal
5. General Bruce Clark
6. "I only regret that I have but one life to lose for my country."
7. Inspector general; training
8. Dennis Hart Mahan
9. Jefferson Davis
10. General Dwight D. Eisenhower
11. Winfield Scott
12. Emory Upton
13. Ulysses S. Grant

14. General Robert E. Lee

15. Harvard Medical School

16. Elihu Root

17. Lieutenant General Matthew B. Ridgeway

18. Captain Merriweather Lewis and Lieutenant William Clark

19. Bridgadier General Anthony Wayne, commander of the Pennsylvania Line

20. British Major John Andre

21. The Battle of Monmouth

22. Major Walter Reed

23. At the First Battle of Bull Run

24. General Douglas MacArthur

25. President (General) Zachary Taylor

26. The Black Hawk War (1832)

27. Captain Charles May to his squadron of dragoons (cavalry) during the Mexican War

28. General James Longstreet

29. Union Major General George H. Thomas and the 19th Infantry

30. Ulysses S. Grant

31. Brigadier General Ulyesses S. Grant during the battle at Fort Donelsan, Tennessee

32. She followed the Union Army during the Civil War, treating the sick and the wounded, and later founded the American Red Cross.

33. Abner Doubleday

34. William T. Sherman on his march to the sea

35. From his initials J.E.B. (James Ewell Brown) Stuart

36. William Tecumseh Sherman. Because they were unable to pronounce his middle name easily, they shortened it.

37. Robert E. Lee

38. Jefferson Davis

39. General Ulysses S. Grant

40. General Francisco "Pancho" Villa

41. Chairman and chief engineer of the Panama Canal Commission and later civil governor of the Panama Canal Zone

42. Lieutenant T.E. Selfridge, who was killed when his Wright plane crashed at Fort Meyer, Virginia in 1908

43. Brigadier General S.L.A. Marshall

44. Lieutenant George S. Patton

45. The 82d Infantry Division

46. Lieutenant William Calley

47. George C. Marshall

48. William "Bill" Mauldin

49. General George C. Marshall to General Dwight D. Eisenhower

50. Washington, Grant, Eisenhower, Taylor and Scott

51. Major General U.S. Grant

52. "I came through and I shall return."

53. General George S. Patton

54. General Mark Clark

55. General "Vinegar Joe" Stillwell

56. Kay Summersby

57. Tokyo Rose

58. General William E. Depuy, commander of the 1st Infantry Division in Vietnam

59. General George S. Patton

60. Lieutenant James N. (Nick) Rowe

61. "Midge," the American-born radio propagandist for the Germans in WWII

62. They were roommates at West Point.

63. Bradley, Clark, Eisenhower, McArthur and Ridgeway

64. In a vehicle accident

65. Bill Mauldin

66. General Omar Bradley

67. Eddie Rickenbacker

68. General Jonathan Wainwright

69. Major General Terry de la Mesa Allen

70. Kilroy ("Kilroy was here.")

71. General William Dean

72. General Walton Walker

73. John O. Marsh Jr., Secretary of the Army (1981-   )

74. Lieutenant Colonel James H. Doolittle

75. General Jonathan Wainwright

76. He is the American Catholic priest who said, "There are no atheists in foxholes."

77. "The Battling Bozos of Bataan"

78. A hand grenade

79. General Lewis B. Hershey, who served for thirty-three years as director of the Selective Service System

80. The developer of K-rations

81. General George S. Patton

82. Captain Colin Kelley

83. General George C. Marshall

84. Italian

85. 35th

86. Colonel William O. Darby (promoted posthumously to brigadier general)

87. General Richard Ewell

88. President Woodrow Wilson

89. Corporal Alvin York

90. A famous water-color artist commissioned to portray the US Army during WWI

91. General John A. Wickham, Jr., Chief of Staff of the Army (June, 1983-   )

92. General James A. Van Fleet

93. He was the top-scoring American air ace of all time, with forty kills.

94. The Army private involved in the celebrated hospital "slapping" incident with General Patton

95. Maxwell Taylor

96. Brigadier General Anthony McAuliffe

97. Kaiser Wilhelm

98. "Crazy Cowboy General"

99. The 37th Tank Battalion, commanded by Lieutenant Colonel Creighton Abrams Jr.

100. George Baker

101. Bob Hope

102. Major General Terry de la Mesa Allen, who commanded the 1st Infantry Division in WWII

103. Colonel George Taylor, commander of the 16th Infantry Regiment, on Normandy Beach

104. A radar specialist, he was the only person who flew on both the Hiroshima and Nagasaki atomic bomb missions.

105. General Creighton Abrams

106. A famous Army pigeon-hero of WWI credited with carrying the message which stopped a barrage of friendly artillery falling on the 388th Infantry, 77th Division

107. She produced the black face cream used as night camouflage.

108. Brigadier General Theodore Roosevelt Jr., assistant division commander of the 4th Infantry Division

109. Colonel Charles Stanton (These words are usually attributed to General Per-

shing; however, on that day he delegated the honor of speaking to Colonel Stanton.)

110. Tojo

111. Twenty-six

112. Kady C. Brownell

113. Sarah Emma Edmundson, a spy for the Union Army in the Civil War

114. He flew C-47's in the China-Burma-India theatre.

115. Lieutenant General Bruce Palmer Jr.

116. George Armstrong Custer

117. General A.P. Hill

118. General Francis Nicholls

119. Captain Lewis L. Millett

120. General Mark Clark

121. A musician (bugler) with Company E, 14th Infantry Regiment. He was awarded the Medal of Honor for being the first individual to scale the walls of Peking during the Boxer Rebellion in 1900.

122. Mary Ann Bickerdyke, a famous Union nurse during the Civil War

123. The Army commander in Hawaii at the time of Pearl Harbor who was held partly responsible because he did not react to warnings of attack

124. General Curtis E. Le May

125. Brigadier General "Billy" Mitchell

126. Brigadier General Douglas MacArthur

127. Colonel Paul W. Tibbetts

128. The first director of the Women's Army Auxililary Corps, and later, as Colonel Oveta Culp Hobby, the first director of the Women's Army Corps

129. Louisa May Alcott

130. General James M. Gavin

131. Human moles

132. Brigadier General Francis T. Dodd

133. General Bruce Clark

134. The leader of the Army's Signal Intelligence Service team that broke the Japanese diplomatic code "Purple" in WWII

135. Lieutenant General Samuel B.M. Young

136. Brigadier General Frederick Davidson, Commander of the 199th Light Infantry Brigade in Vietnam

137. Captain Roger C. Donlon

138. Joyce Kilmer

139. Major Thomas D. Howie, Commander of the 3d Battalion, 116th Infantry, 29th Infantry Division

140. General John J. Pershing

141. General Bruce Palmer Jr.

142. Alexander Haig

ARMY LORE

1. 1959
2. The kepi-type hat used in the Civil War
3. A long spear-like weapon used by NCO's in Revolutionary War times to direct soldiers and to keep the lines
4. Wood
5. A .75-caliber musket used during the Revolutionary War
6. The whiskey ration
7. The railroad, the telegraph and the rifle
8. 7.91 pounds
9. First, confusion of uniform colors led to the adoption of gray for Confederate troops and blue for Federal troops. Second, the Confederate battle flag was changed to make it more distinguishable from the Union flag.
10. A tourniquet has been placed on his wound.

11. Survival, Evasion, Resistance and Escape

12. Military Occupational Specialty

13. Forward Line of Own Troops (FEBA was the Forward Edge of the Battle Area.)

14. Six

15. It is the rule for the placement of enlisted men's brass on their uniform.

16. Army Training and Evaluation Program

17. Cold

18. The Spanish-American War

19. A mule-drawn wagon used by officers of high rank to carry their baggage and equipment in the mid-1850s

20. The camel

21. An alert status, usually ordered before first light, during which a combat unit prepares itself for an impending attack

22. Harassing and Interdictory fires

23. Armed Personnel Carrier

24. "I will guard everything within the limits of my post and quit my post only when properly relieved."

25. An orderly room

26. Guard mount

27. A rope-like device worn on the arm to signify award of certain foreign decorations

28. Lead

29. 1,100 meters

30. Close Station, March Order

31. 3,600

32. The standard US Army artillery piece in WWI

33. A device used to emplace ("lay") artillery and mortars

34. Adjutant's call

35. An OCS graduate

36. Escort the unit colors

37. A drill sergeant

38. A Fire Direction Center

39. An emergency artillery or mortar fire mission called while a firing unit is on the move

40. Blue

41. Driving at night

42. The small flag carried by a company-sized unit

43. A coaxially-mounted machine gun located on the turret of an armed vehicle

44. Left

45. Absent With Out Leave

46. The poncho

47. A mortar

48. The operations officer (S-3/G-3/J-3)

49. Sabre

50. 1. Situation; 2. Mission; 3. Execution; 4. Logistics; and 5. Command and Signal

51. A bullet-proof vest

52. A helmet

53. 40 mm

54. "Gun"

55. The Chinook (CH-47)

56. That the wearer is a combat leader

57. Fragmentation, Riot Control, Smoke, and Incendiary

58. A prescribed period usually at the beginning of the day during which all soldiers in need of medical attention are treated

59. False

60. Rear

61. Right and up

62. One

63. The insignia of the Army Medical Corps

64. Left

65. An organized period of vehicular maintenance

66. WWI

67. A hand grenade used in WWI

68. The duty roster

69. "Halt! Who goes there?"

70. Right

71. Left

72. Tucked into or secured at the top of the boot

73. Olive blouse and gray trousers

74. Late afternoon/early evening

75. Morning

76. Daily

77. A technical sergeant (E-5)

78. Immediate action to reduce a stoppage on an M16A1 rifle

79. Heat stroke, heat exhaustion and heat cramps

80. "Password"

81. Left

82. An artillery weapon

83. A sort of cart that was placed between the team and the artillery piece in the horse-drawn artillery era

84. A sergeant, E-5, three stripes

85. Two

86. The command sergeant-major

87. First sergeant

88. Forward Observer

89. "To the Colors" and "Retreat"

90. A platoon

91. A second lieutenant

92. A 2½-ton truck

93. Tube-launched, Optically-sighted, Wire-guided

94. A contemptuous term applied during WWII to conscientious objectors

95. Light Antitank Weapon

96. Greenwich Mean Time

97. The General Orders of a Sentry

98. England

99. "Huey (UH-1)" and "Blackhawk (UH-60)"

100. The Cobra (AH-1)

101. Magnetic azimuth

102. **S**ize
Activity
Location
Unit
Time
Equipment

103. **W**idth (in meters)
Over
Range (times)
Mils (in thousands)

104. Sense and Destroy Armor

105. An antiarmor missile system developed for use with the attack helicopter

106. Front Line Ambulance

107. The opener used for cans of field rations

108. A Remotely-Piloted Vehicle

109. The "Locust"

110. Ten

111. Direction

112. A Final Protective Line

113. 875 mils

114. The Rocket Launcher, T-34

115. 1) Hill
2) Ridge
3) Valley
4) Saddle
5) Depression

116. The Mine Exploder, T1E3 (M1)

117. Yes

118. The absence of any obstruction in the path of the trajectory of a weapon

119. Beginning of Morning Nautical Twilight, (i.e., daylight)

120. A helicopter equipped with searchlights to illuminate ground targets

121. Two

122. Leapfrogging

123. A device used to Traverse and Elevate a machine gun

124. Mission-Oriented Protective Posture

125. Armored Combat Earthmover

126. A Fire Support Team

127. Yes

128. Cocked hat

129. Night defensive positions

130. Key terrain

131. Article 15

132. 1) Summary
2) Special
3) General

133. An anti-personnel mine

134. Three

135. Fire delivered to sweep the length of a line of troops, a trench or other targets

136. Airmobile extraction

137. Armpit deep

138. Nine seconds

139. Every half-hour

140. One gun for each state

141. The Northrop P-61 fighter

142. None

143. A six-wheel-drive amphibious truck used in WWII

144. A 150-mm gun placed on a Sherman tank

145. A 76-mm tank destroyer

146. The bazooka

147. The "Flying Fortress"

148. "Eagle Flight"

149. One

150. 1) Garrison
2) Post
3) Field
4) Storm

151. L-4 and L-5 observation and reconnaissance aircraft used during WWII

152. The General Stuart Tank

153. The Model 1903 Springfield rifle

154. None. Although they are usually referred to as "crossed rifles," the official infantry insignia consists of crossed, 1795 model, Springfield Arsenal muskets.

155. It indicates the rank of the general officer for whom the aide works.

156. Thirteen

157. Protection from fire or observation by an obstacle such as a ridge, a hill or a bank

158. An area within the maximum range of a weapon, radar or an observer which cannot be covered by fire or observation

159. Artillery, tank and recoilless rifle ammunition containing hundreds of small steel flechettes. It is used in an anti-personnel role.

160. Twin 40-mm cannon mounted on a tank chassis, originally an anti-aircraft weapon

161. The nickname for the WWI hand grenade

162. The flame-thrower

163. The largest US artillery weapon of WWI

164. A "doughboy"

165. The "Liberator"

166. The B-23 bomber

167. A term, originating during the Vietnam War, for a UH-1 helicopter

168. A medical evacuation mission

169. End of Evening Nautical Twilight (i.e., dusk)

170. A commercial air blower used to blow smoke into tunnels in Vietnam in order to identify the many exits the tunnels usually contained

171. Army air defense missiles

172. The antimalarial drug administered during WWII

173. Twenty-one

174. The B-25

175. The Presidential Unit Citation

176. The president

177. In precedence: Presidential Unit Citation, Valorous Unit Award, Joint Meritorious Unit Award and Meritorious Unit Commendation

178. The "Marauder"

179. A six-inch gun that was one of the primary weapons used by the Coast Artillery for harbor defense

180. A bulldozer with a sharp blade capable of felling trees

181. No

182. "Arclight"

183. Barbed wire

184. Distinguished Service Cross

185. Master Parachutist, Senior Parachutist and Parachutist

186. Indian Campaign Medal

187. United States Army

188. 65%

189. Combat

190. A shoulder tab awarded to the top 100 successful contestants in the President's Match, held annually at the National Rifle Match

191. Cadets, United States Military Academy

192. A "Duster"

193. The NCO member of a field artillery Forward Observer team

194. A Tactical Operations Center

195. A rope-like device worn on the left shoulder signifiying that the wearer is an aide-de-camp to a general officer.

196. Prescribed Load Lists

197. Authorized Stockage List

198. Tactical Area Of Responsibility

199. Area of Operations

200. Long Range Reconnaissance Patrol

201. A converted DC-3 transport used in Vietnam. It was equipped with rapid-fire machine guns capable of firing thousands of rounds per minute

202. Salutes

203. Army teams which are dropped or air-landed in advance of a parachute or air-mobile operation for the purpose of guiding aircraft to the drop/landing zones

204. Flight as close to the earth's surface as possible, following the contours of the earth

205. "Lift"

206. A Rear Area Operations Center

207. The M-113 series of armored personnel carriers

PHOTOGRAPHS

1. Lieutenant General Joseph W. Stilwell
2. Sixty-one
3. 1st Battalion, 3d Infantry
4. The Old Guard
5. General George C. Marshall
6. 82d Airborne Division
7. Fort Hood, Texas
8. Epinal American Cemetery in France
9. Lieutenant General George S. Patton Jr.
10. Sicily
11. It is the main entrance to McGregor Guided Missile Range.
12. Fort Bliss, Texas
13. Major General William C. Westmoreland

14. West Point cadets

15. Fort Benjamin Harrison, Indiana

16. Lieutenant General Matthew B. Ridgeway, Major General Doyle O. Hickey and General of the Army Douglas MacArthur

17. Fort Benning, Georgia

18. General Maxwell D. Taylor and Major General William C. Westmoreland

19. Fort Campbell, Kentucky

20. The top medal

21. 1862

22. General Mark W. Clark

23. Post Headquarters, Fort Devens, Massachusetts

24. Fort Leonard Wood, Missouri

25. Fort Ord, California

26. **Cavalry Charge**

27. Washington, DC

28. The Washington Monument

29. United States Military Academy at West Point

30. Fort Leavenworth, Kansas

31. Abilene, Kansas

32. General of the Army
33. Fort Lewis, Washington
34. The Army War College
35. Fort McNair
36. Presidio of San Francisco, California
37. First Division Monument
38. Washington, DC
39. Fort Riley, Kansas
40. The 1st Infantry Division
41. Quarters Number One
42. Fort Myer, Virginia
43. West Point
44. The library
45. **The Infantryman**
46. Fort Benning, Georgia
47. General Roscoe Robinson, Jr.
48. United States Military Academy
49. A "battle sled," a field-expedient designed to protect soldiers advancing over open beaches
50. World War II
51. **The Airborne Trooper**
52. Fort Bragg, North Carolina

ARMY HISTORY

1. "The Army Goes Rolling Along"
2. June 14, 1775
3. The 182d Infantry Regiment and the 101st Engineer Battalion of the Massachusetts Army National Guard
4. The British evacuated Boston.
5. "Sentiments on a Peace Establishment"
6. Battery D, 5th Artillery Battalion, 1st Infantry Division
7. Sylvanus Thayer
8. The Union Army in the Civil War—or the Union Army Veterans Organization
9. Aviation
10. The 1st Division
11. A force of 10,000 enlisted volunteers who possessed immunity to tropical diseases. They were formed into ten infantry regiments during the Spanish-American War.

12. The Dick Act
13. The Manchu Law
14. Army Air Corps
15. The Department of Defense
16. Militia
17. "This we'll defend."
18. 168
19. The signing of the peace treaty, the Treaty of Ghent, ending the war
20. Dr. Anita Newcomb McGee
21. She was the only woman ever to receive the Medal of Honor.
22. The Mexican War
23. Thirteen days
24. The Code of Conduct for members of the United States Armed Forces
25. The Grand Division
26. The 9th Infantry Regiment
27. The Militia Act of 1903
28. Aviation
29. Eight years
30. The first division-size parachute operation in the history of the US Army by

the 504th and 505th Parachute Infantry Regiments of the 82d Airborne Division near Gela, Sicily

31. The US Army Corps of Engineers

32. June 14, 1956

33. The Grand Union flag

34. Department of the Army

35. The 13th Infantry

36. The 28th Infantry Regiment

37. The Active Army, the Army Reserve Components (USAR and ARNG) and the Army's Civilian Workforce

38. Regular Army, National Guard and National Army

39. 1967

40. May, 1950

41. 1959

42. Department of Defense

43. 1946

44. 1923

45. President Wilson's "Fourteen Points"

46. 1802

47. Commanding General of the Army

48. Casablanca

49. The Lend Lease Act

50. The **Enola Gay**

51. The draft

52. 1969

53. 1942

54. The 100th Training Division

55. Lieutenant General Jonathan M. Wainwright

56. Khaki

57. April 23, 1908

58. John C. Calhoun

59. Massachusetts, Connecticut and Rhode Island

60. The 27th Volunteer Infantry Regiment of Maine. However, a Congressional board of review in 1916 found the medals to have been improperly awarded and the recipients' names were stricken from the Medal of Honor list.

61. The "Rough Riders"

62. It was an all-black squadron.

63. Lieutenant General Lesley J. McNair

64. "Mickey Mouse"

65. The name of Army forces committed to Cuba during the Spanish-American War

66. John J. Pershing, Peyton March and Tasker Bliss

67. 1952

68. The battle at Fort Ticonderoga on May 10, 1775

69. Ten

70. Three

71. Chief of Staff of the Army

72. Three

73. The practice of flogging

74. During the Navajo campaign of 1860

75. 1821

76. The British burned the Capitol and the White House.

77. July 29, 1775

78. 1940

79. 1967

80. The collective name of the 138 "colored" regiments of infantry organized by the Union during the Civil War

81. 1918

82. Name, rank and unit

83. True

84. The "Ike" jacket

85. November 17, 1775

86. Yes

87. September 26, 1914

88. March 11, 1779

89. Fourteen

90. Blue and gray

91. Twenty-five

92. He is called the father of American airborne forces.

93. Henry Cabot Lodge

94. The 27th and 31st Infantry Regiments

95. Ten

96. The formal title of the "Boxers" of the Boxer Rebellion

97. February 23, 1980

98. 1920

99. The United States Army Corps of Engineers

100. Olive drab

101. Private Eddie D. Slovik

102. The Malmédy Massacre

103. Six

104. Eleven

105. Thirteen

106. Cairo

107. Potsdam

108. Thirty-eight

109. Ten

110. Eight

111. Split Command

112. "Military Order of the Cootie"

113. VE Day (Germany surrendered.)

114. The **Dorchester**

115. The North Korean invasion of South Korea

116. The 9th Armored Division

117. Compiégne

118. Four

119. "Yorktown 1781"

120. Yes. First Lieutenant Sharon A. Lane of the Army Nurse Corps was the only female servicewoman killed as a result of enemy action in Vietnam.

121. September, 1947

122. The Americal Division and the Phillipine Division

123. The Civil War (corps cloth badges)

124. 1954

125. Cairo, Egypt

126. Combat Commands

127. 60 to 65 miles

128. The name given to the First Special Service Force, a unit composed of Canadians and Americans

129. Racial integration

130. It was made up of Norwegian-Americans.

131. They were assigned to the western frontier and given missions that did not involve fighting against the Confederacy.

132. It took effect at 11:00 a.m. on November 11, 1918.

133. The National Defense Act of 1916

134. June 14, 1775

135. They were all-black divisions. (Many units had white officers, however.)

136. The Axis

137. One

138. Seventeen

139. The US Army Corps of Engineers

140. **The Stars and Stripes**

141. October 25, 1983

142. Civil War, Korean War and Vietnam War

143. During the Berlin Crisis in the summer of 1961

144. XVIII Airborne Corps

145. To protect the lives of American nationals and escort them home

146. The US, Brazil, Costa Rica, Honduras, Nicaragua and Paraguay

147. Australia

148. Three (two Army National Guard and one Army Reserve)

149. Eight

150. The Howze Board

151. Sixty-four

152. Combat Arms Regimental System

153. The US Aerial Mail Service

154. The WWII plan for the invasion of Japan. It was never executed due to the Japanese surrender.

155. Family members

156. Family members killed

GENERAL

1. Lieutenant (later Brigadier General) Edmund L. Gruber

2. Fort Monmouth, New Jersey

3. The Whiskey Rebellion

4. United States Military Academy

5. The brevet system

6. Washington DC or Monterey, California

7. Carlisle Barracks, Pennsylvania

8. Substitution (paying another to perform one's military service) and commutation (purchasing outright relief from military service)

9. The Post Exchange

10. The 3d Infantry Division

11. William Tecumseh Sherman

12. In the 17th century British Army, the senior grade was a lieutenant general.

The next-in-command was called a "sergeant-major-general." In the course of the intervening centuries, the "sergeant" has been dropped from usage.

13. Fort Gordon, Georgia

14. The 7th Cavalry

15. Ninety-one

16. Triangular

17. Aberdeen Proving Grounds, Maryland

18. American Expeditionary Force

19. The Purple Heart

20. Chief Tecumseh of the Shawnees, who had been commissioned a brigadier general in the British Army

21. Springfield Arsenal

22. Fort Knox, Kentucky

23. The 4th Infantry Division

24. "Taps"

25. The Medal of Honor

26. Traveler

27. German gasses used against the US Army in WWI

28. A concept originated by President Theodore Roosevelt to form a division of volunteers to be sent to Europe in WWI. It was not approved.

29. An ointment used in WWI to prevent and treat burns received from mustard gas

30. Fort Sam Houston, Texas

31. The 82d Airborne Division

32. An organization created by Congress after the Civil War and administered by the US Army. It assisted former slaves with food, clothing, employment, homesteads, etc.

33. The demilitarized zone between North and South Korea

34. The 4th Armored Division

35. Fort McLellan, Alabama

36. Personnel and administration

37. Fort Benning, Georgia

38. The 7th Cavalry

39. One hundred

40. A procedure wherein a promotable person is awarded his rank in advance, but does not receive the higher pay

41. Marine

42. The artillery

43. Organization Day

44. "Ruffles and Flourishes"

45. Three

46. 25th

47. A major general

48. Saint Barbara

49. Forward

50. The 7th Infantry Division

51. Yellow

52. Fort Sill, Oklahoma

53. Mobilization

54. North American Air Defense

55. A corps

56. Heavy (or mechanized or armored), airborne, light and airmobile

57. Orange

58. The Land Grant Act of 1862

59. 1. Valor
    2. Achievement
    3. Meritorious service

60. The Judge Advocate General School

61. "Rodger Young"

62. The Transportation School

63. The 101st Airborne Division

64. "Follow me."

65. The 187th Airborne Regimental Combat Team

66. Command Sergeant-Major of the Army

67. Battalion/squadron-size

68. The infantry

69. Lieutenant colonel

70. Non-commissioned and warrant

71. Company/battery/troop-size

72. A civilian camp follower ("military peddlar") who had permission to sell such things as food, liquor, tobacco and other items, impossible to obtain otherwise, to Army troops in remote garrisons and on the frontier

73. 1,570,000

74. The 26th Infantry

75. The 24th Infantry Division

76. Airborne, mechanized, airmobile, and light

77. 1. c
    2. d
    3. a
    4. e
    5. b

78. The 42d Infantry Division

79. The Flying Tigers
80. Brigade or group-size
81. Fort Bliss, Texas
82. Fort McClellan, Alabama
83. P-61
84. Fort Benning, Georgia
85. The US Army Aviation Center and School
86. The US Army War College
87. 1946
88. The 2d Armored Division
89. Four million men
90. The 1st Cavalry Division
91. After the Battle of Cantigny in France in 1918
92. Fort Lee, Virginia
93. Sergeant Barry Sadler
94. Fort Benjamin Harrison, Indiana
95. Seventeen and twenty-two
96. The 3rd Armored Division
97. The Army Intelligence School
98. 1942

99. The 5th Infantry Division (mechanized)

100. Fort Bragg, North Carolina

101. 5,986,000

102. Fort Leavenworth, Kansas

103. The 2d Infantry Division

104. It placed the Army Corps of Engineers in control of the nation's seaports and inland waterways.

105. The 34th Infantry Division

106. The 36th Infantry Division

107. Unlimited

108. The "Buckeye Division"

109. The 38th Infantry Division

110. Five

111. A name given to Avenger Field, located near Sweetwater, Texas—home base of the WWII WASP's

112. The "Sunburst Division"

113. "No Expense"

114. Norman Rockwell

115. The GI Bill

116. Airborne

117. The 29th Infantry Division

118. The "Thunderbird Division"

119. The "Dixie Division"

120. Fraunces Tavern

121. The 63d Infantry Division

122. Five

123. The 30th Infantry Division

124. 1st Cavalry Division

125. Thirteen

126. Field Artillery

127. The Women's Army Corps

128. The Regiment of Mounted Rifles, today's 3d Armored Cavalry Regiment

129. 1980

130. The Golden Knights

131. "Old Ironsides"

132. An association formed by a group of Revolutionary War officers to perpetuate their mutual friendships and assist officers and their families who were in need of help

133. "Valor"

134. The 11th Airborne Division

135. Female physical and occupational therapists employed by the Army in WWI

136. No. They did serve the Army, however, as civilians under contract.

137. The 76th Infantry Division

138. The Greek goddess Pallas Athena

139. Women Airforces Service Pilots, an organization of about one thousand women who were civilian employees of the Army during WWII; they served as ferry pilots, mechanics, instrument instructors, etc.

140. The 70th Infantry Division

141. The "Spearhead Division"

142. The 4th Armored Division

143. The 77th Infantry Division

144. Forty-three

145. The 78th Infantry Division

146. Sixty-two

147. The "Lorraine Division"

148. They were raided by the forces of General Francisco (Pancho) Villa on March 9, 1916.

149. Winfield Scott's strategic concept for the conduct of the Civil War. It was not adopted, but came about anyway, three years after he proposed it.

150. 1946

151. The US Army Pigeon Service

152. The Signal Corps

153. The "Wildcat Division"

154. On January 5, 1777 kegs of powder were floated on the Delaware River to destroy ships of the British fleet.

155. "Tunnel rats

156. A device used by certain paratroopers that enabled them to carry a carrier pigeon with them on jumps

157. "Huey"

158. Twelve

159. It was the airfield on which the 82d Airborne Division landed during the Dominican Crisis in 1965.

160. "Blue Ridge Division"

161. Forward Air Controllers (FAC's)

162. A soldier trained and qualified to handle Army pigeons

163. The 32d Infantry Division and the 49th Armored Division

164. Return of Forces to Germany

165. The Women's Army Auxiliary Corps (WAAC)

166. Armor

167. "Thunderbolt"

168. 1975

169. The form of organization of the Army directed by the president and Congress in 1792. It was composed of four "sub-legions," each having two battalions of infantry, one battalion of rifles, a troop of dragoons and a company of artillery.

170. The 81st Infantry Division

171. 26th Infantry Division

172. The Women's Army Corps (WAC)

173. The 84th Infantry Division

174. The Order of Finfinella

175. The "Black Hawk Division"

176. In 1815 the War Department ordered that it thereafter be of the gray color worn by Winfield Scott's brigade during the Battle of Chippewa in July, 1814.

177. The Chemical Corps

178. The 88th Infantry Division

179. Because the letter "J" was too easily confused with the written letter "I."

180. The 8th Infantry Division

181. 3,700 pounds (a pilot, co-pilot and thirteen fully-equipped soldiers)

182. The three chords played by an Army band before and after it troops the line

183. The 12th Armored Division

184. It is the instrument used to legalize the authority of an officer who must be appointed to a position of command over an officer senior to him.

185. The first "squads," which were organized in 1855 as four-man teams

186. Military Police

187. 1905

188. The 89th Infantry Division

189. 1976

190. A badge or lapel pin worn above the right breast pocket to indicate that the wearer had served honorably overseas in WWII and was heading home

191. "Tough 'Ombres"

192. The name affectionately given to female USO employees in Vietnam

193. 1968

194. **The Bell**

195. 92d

196. The words literally mean "open arms." It was a program under which the South Vietnamese government offered amnesty to Viet Cong defectors.

197. "Constructive Months Service." It was the point system used to determine eligibility for rotation from Korea.

198. The 91st Infantry Division

199. Irving Berlin

200. The 95th Infantry Division

201. Company/battery/troop

202. 96th

203. The Army Nurse Corps

204. The 32d Infantry Division

205. The 98th Infantry Division

206. Five

207. 97th

208. Military Intelligence

209. Fort Monmouth, New Jersey

210. Air Defense Command

211. Roman

212. The US Army infantryman

213. A classified study called "The Program for the Pacification and Long-Term Development of South Vietnam," which was commissioned in 1965 by Army Chief of Staff Harold K. Johnson to look in the broadest possible way at the situation in South Vietnam

214. A Studies and Observations Group established in 1964 to conduct US clandestine unconventional warfare, initially in North Vietnam. Later it expanded to Cambodia and Laos.

215. Forty-three

216. Point Salinas Airfield, Grenada

217. The 100th Infantry Division

218. Libby Prison, near Richmond, Virginia

219. 102d

220. 100

221. Chief Commander, Commander, Officer and Legionnaire

222. A device wherein each Medal of Honor awardee, who so desires, is certified by the Veteran's Administration as being entitled to receive a special pension of $200.00 per month

223. The name given by WWII GI's to powdered milk ("milk" spelled backwards)

224. One of the B-25 bombers in the Doolittle Raid piloted by Captain Ted Lawson who later wrote **Thirty Seconds Over Tokyo**

225. A nickname for the Ledo Road, built by Lieutenant General Lewis A. Pick

226. "Benzine Boards"

227. Perforated steel planks used to surface airfields in WWII

228. Army slang for a chaplain

229. "Lightning"

230. The Phillipine Scouts

231. Yellow fever, which was conquered by Army doctors during the Spanish-American War

232. Rear

233. "Operation Little Switch"

234. Willie

235. The Italian propaganda broadcaster who used the title "Axis Sally"

236. The WWII fighter plane commander who was the model for the cartoon character "Flip Corkin"

237. A WWII poster character used in occupied Germany. Her initials warned American soldier of the dangers of fraternization.

238. **Private Hargrove**

239. "Mustang"

240. The final prisoner exchange of the Korean War

241. "Timberwolves"

242. The Army Mutual Aid Association

243. The Bantam Car Company

244. Yes; however, it was later taken away.

245. An award given to the outstanding Army ROTC graduate each year

246. Army Staff Identification Badge

247. Minerva, the Roman goddess of wisdom and righteous war

248. The 1st Armored Division